MIRACULOUS MEDICINES AND THE CHEMISTRY OF DRUG DESIGN

Global Science Education
Professor Ali Eftekhari
Series Editor

Learning about the scientific education systems in the global context is of utmost importance now for two reasons. Firstly, the academic community is now international. It is no longer limited to top universities, as the mobility of staff and students is very common even in remote places. Secondly, education systems need to continually evolve in order to cope with the market demand. Contrary to the past when the pioneering countries were the most innovative ones, now emerging economies are more eager to push the boundaries of innovative education. Here, an overall picture of the whole field is provided. Moreover, the entire collection is indeed an encyclopaedia of science education and can be used as a resource for global education.

Series List:

The Whys of a Scientific Life
John R. Helliwell

Advancing Professional Development through CPE in Public Health
Ira Nurmala and Yashwant Pathak

A Spotlight on the History of Ancient Egyptian Medicine
Ibrahim M. Eltorai

Scientific Misconduct Training Workbook
John Gaetano D'Angelo

The Whats of a Scientific Life
John R. Helliwell

Inquiry-Based Science Education
Robyn M. Gillies

Hark, Hark! Hear the Story of a Science Educator
Jazlin Ebenezer

Miraculous Medicines and the Chemistry of Drug Design
Nathan Keighley

The Whens and Wheres of a Scientific Life
John R. Helliwell

MIRACULOUS MEDICINES AND THE CHEMISTRY OF DRUG DESIGN

Nathan Keighley

CRC Press
Taylor & Francis Group
Boca Raton London New York

CRC Press is an imprint of the
Taylor & Francis Group, an **informa** business

First edition published 2021
by CRC Press
6000 Broken Sound Parkway NW, Suite 300, Boca Raton, FL 33487-2742

and by CRC Press
2 Park Square, Milton Park, Abingdon, Oxon, OX14 4RN

© 2021 Taylor & Francis Group, LLC

CRC Press is an imprint of Taylor & Francis Group, LLC

Library of Congress Cataloging-in-Publication Data

ISBN: 978-0-367-64403-1 (hbk)
ISBN: 978-0-367-64407-9 (pbk)
ISBN: 978-1-003-12439-9 (ebk)

DOI: 10.1201/9781003124399

Typeset in Times New Roman
by MPS Limited, Dehradun

Contents

Preface

The subject of chemistry is widely acknowledged as being conceptually challenging, and regarded with a perceived elitism. Consequently at school, many individuals are put-off by chemistry and tend to avoid it in later education. For me, this is unfortunate because chemistry is a beautiful, interesting, and exceedingly important and relevant subject; playing an essential role in society. This manuscript aims to enlighten the reader to how important chemistry is in terms of medicine. Of course, chemistry's role in society is by no means limited to medicine: materials, food, and many technologies rely of chemists and their passion and intrigue for the subject; without which there would be severe limitations to the advancements of these technologies.

At its core, chemistry is a rigorous and systematic subject. This is very much reflected in the nature of chemistry textbooks used in education. However, many people might be put-off by this format and the interesting aspects of the subject may be lost. This manuscript hopes to address this dilemma by producing a text that is a joy to read and makes the subject interesting while still including the foundational principles of chemistry, and not lose the systematic rigour of the subject.

The systematic nature of chemistry, for me, made it a delight to learn because my studies had structure. In particular, I found organic chemistry to be fascinating, and the main goal for this manuscript was to convey my enthusiasm for organic chemistry as well as highlight its fundamental importance in medicine. Miraculous medicines aims to be an introductory text, so that readers with any, or no, background education in chemistry can access and enjoy the material and gain an appreciation for the subject, which is intertwined in all our lives.

A Brief Background

1

Life is based on carbon. Organic chemistry is dedicated to this element, whose properties are defined by the nature of the carbon atom. Carbon is in group 4 of the periodic table, which identifies that there are four valence electrons. These are the outer electrons involved in bonding, so a further four electrons are required through bonding to other atoms to satisfy the octet rule; hence, carbon atoms characteristically form four covalent bonds, and each bond comprise of shared pairs of electrons with partner atoms. These may be single bonds, with one shared pair of electrons, or double bonds and even triple bonds, with two and three pairs of electrons being shared. It is common for carbon atoms to bond together in chains and rings to produce the carbon skeleton that defines an organic molecule; the remaining valences are often satisfied by hydrogen atoms, which offer their single electron to form a bonding pair. This produces the simplest class of organic compounds: hydrocarbons, typically separated according to molecular mass through fractional distillation of crude oil.

Other elements from the periodic table, generalised as heteroatoms, are also of importance in organic chemistry. Particularly, nitrogen, oxygen, and the halogens (group 7), and even organometallic species are all vital components of organic chemistry. To fulfil the octet rule, nitrogen, with five valence electrons, requires three more electrons, so typically it forms three bonds; oxygen (group 6) needs two electrons, so tends to form two bonds; while the halogens require one more electron, so form single bonds. The valence electrons govern the chemistry of the elements, and so are the most important. However, they do not represent the entirety of the atoms electron configuration; it is worth noting that the total number of electrons, whose charge is fundamentally negative, equals the atomic number, Z, which is the number of positively charged protons within the atomic nucleus to give a neutral atom. The remainder of the atomic mass, which for carbon is 12 g mol^{-1}, is made up with neutrons in other words the carbon nucleus is composed of six protons and six neutrons. The number of neutrons may vary to give isotopes, which have a different atomic mass but the same chemical properties, since it is the valence electrons that govern chemistry. The appropriate number of valence electrons may be added or subtracted to give charged atoms, called ions.

The number of covalent bonds that an atom forms governs molecular geometry. This is explained by valence-shell electron pair repulsion (VSEPR) theory, which predicts that pairs of electrons will occupy positions around the atom as far apart from one another as possible. This leads to characteristic molecular shapes. For a carbon atom within a molecule, with four equivalent bonding pairs of electrons, each bonding pair will be positioned equidistantly in three-dimensional space to produce a tetrahedral geometry, with equal 109.5 degree angles between bonds. For a boron atom forming three equivalent bonds, the bonding pairs of electrons will be separated the maximum distance apart to produce a trigonal planar shape, with 120 degree bond angles. Scenarios where multiple bonding is present result in different molecular shapes. If the carbon atom was to have a double bond with one of its neighbours, as in the case of ethene, there will be three points of electron density and, consequently, the molecule will adopt a trigonal planar shape. However, the electron density is not evenly distributed around the carbon atom; more of it is present in the double bond than in the two single bonds, and, therefore, the shape will be slightly distorted. VSEPR theory predicts that electrons that are not involved in bonding, or 'lone pairs', will repel more and influence molecular shape. For example, although a nitrogen may have three bonds similar to boron, the presence of a lone pair pushes the bonding electron pairs away to produce a trigonal pyramidal shape. Likewise, water molecules are not linear; the presence of two lone pairs pushes the oxygen-hydrogen bonds down into a V-shape. The shapes of molecules, dependent on the nature of the atoms and their bonding, have a dramatic effect on chemistry because molecular geometry influences how reactions proceed.

Reactions of organic molecules are, for the most part, governed by the presence of heteroatoms. They have the ability to disturb the electron density within the local area of the hydrocarbon skeleton and therefore create a reactive centre. The positioning and nature of the bonding of heteroatoms in organic molecules are identified as functional groups, which will undergo characteristic reactions.

Understanding electrons is essential to chemistry. In a reaction, chemical bonds must be broken: this may be a heterolytic cleavage, where two electrons in the bond move to one species to form ions, or a homolytic cleavage, where the pair of electrons are shared to produce free radicals. In organic chemistry, the movement of electrons is shown with curly arrows to produce organic reaction mechanisms, which will feature later in the text. Since reactivity is the movement of electrons to break weak bonds and make new, stronger bonds, it is possible to predict how an organic reaction mechanism will proceed. For two reacting molecules, identify where the electrons are coming from. This molecule is termed the nucleophile—a negatively charged ion, or neutral molecule with a lone pair of electrons which are donated to form a covalent bond. The electrons are received by the electron-deficient molecule called an electrophile. Whether a given molecule will react as a nucleophile or an electrophile depends on the functional groups that are present.

Organic reactions can be classified as either acid-base reactions or redox reactions. The transfer of a hydrogen ion (a proton) identifies an acid-base reaction, while a change in functional group from reactants to products shows reduction-oxidation reactions. Processes that involve homolytic bond cleavage are called radical reactions. Processes that involve heterolytic bond breaking are called polar reactions. Polar reactions are the most common type of reaction of organic molecules and involve reactions between polar molecules and/or ions. There are three main classes of polar reactions. An addition reaction involves the combining of two molecules to yield a single product, while in an elimination reaction, one reactant molecule is converted into two product molecules. In a substitution reaction, one functional group on the molecule is replaced by another.

Two organic molecules can be composed of the exact same proportions of atoms in other words have the same molecular formula, but the way in which the atoms are arranged may be completely different to give two unique molecules. These are termed isomers. There are three different types of structural isomerism. The functional group may be placed at different positions on the carbon chain in positional isomers, or the atoms may be arranged in such a way to give a different functional group, known as functional group isomerism. Equally, it can be the hydrocarbon chain itself that is arranged differently to give different chain isomers.

Figure 1.1 demonstrates how atoms can be arranged to give different molecules from the same molecular formula. Note that the way in which these molecules are drawn is the skeletal formula. Each corner represents a carbon atom; only important heteroatoms are labelled and the remaining valences of the carbon atoms are bonds to hydrogen atoms (not drawn to save convolution). It can be clearly seen that three of the molecules are alcohols (OH) and molecule (c) is a different functional group. The position of the OH is

FIGURE 1.1 Structural Isomers Possible for the Molecular Formula $C_4H_{10}O$.

different between molecules (a) and (b), while in molecule (c), the carbon chain has branched. Each of these molecules will have different properties and reactivity.

The way in which these molecules are drawn does not portray their three-dimensional structure, which is important because molecules with the same structural formula can be arranged differently in space to produce stereo-isomerism. Molecules are continuously moving and vibrating and rotation around single bonds, which means that they can adopt different conformations. Molecules with the same structure can also exhibit different configurations, where they may exist as non-superimposable mirror images, or groups of atoms may be held in different spatial arrangements on either side of a rigid carbon-carbon double bond. This concept may be difficult to envisage, but is crucially important in drug design, as will be seen later in the text. For example, the unfortunate consequences of thalidomide, used for morning sickness, was due to the drug been administered as a fifty-fifty mixture of the mirror images, where one of the spatial arrangements caused harm.

Organic reactions are used to synthesise drugs. Considerations of the stereochemistry are obviously vital in the design of new medicines. In cases where one of the two stereoisomers is the active drug, an asymmetric synthesis is required where special measures are taken to ensure stereo-specificity. With knowledge of the characteristic reactions that different functional groups display, organic chemists can synthesise a target drug molecule from the relevant readily available starting materials. To build a target molecule, making carbon-carbon bonds is essential. Functional groups that will undergo addition reactions are useful for this purpose. To ensure strong interactions with the drugs' biological target, a particular functional group may be needed in the molecule. Here, a substitution reaction may be relevant. Ultimately, a drug molecule is made with the correct size, shape, correctly positioned functional groups, and chemical properties that will interact with the biological system to produce a biological response. A selection of functional groups that are key to organic synthesis are shown in Figure 1.2. Whether the compound acts as a medicine or a poison depends on the dose level of the compound. This can be described by the drugs therapeutic index, which is a measure of a drugs beneficial effect at low dose versus its harmful effects at high dose. No drug is absolutely harmless and drugs may vary in the side effects they have.

The design of a medicinally useful compound is a long and arduous process. The first difficulty is identifying a biological target for which a therapeutic molecule can be designed that will interact in such a way as to combat the disease. Understanding the molecular basis of disease is paramount in order to be able to design a compound with the correct chemical structure to not only bind to the biological target, but also interact in a way

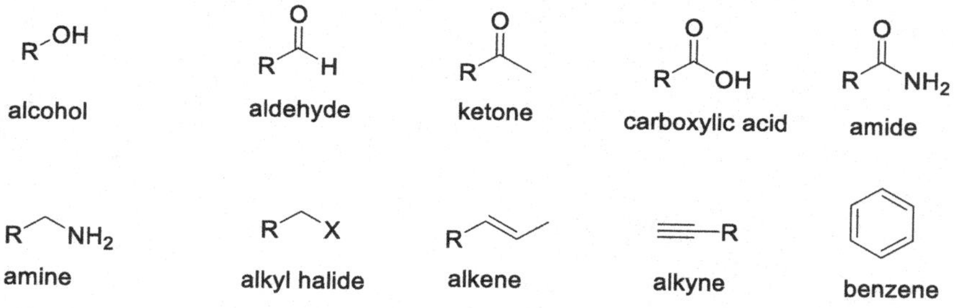

FIGURE 1.2 A List of Selected Functional Groups Where R = the Rest of a Molecule and X = Halogen Atoms in other words F, Cl, Br, and I.

that produces a biological response that suppresses the malfunctions caused by the disease that have an adverse effect on good health. Once a target is identified and a target molecule determined through computational analysis, organic chemists strive to synthesise this lead compound. Many slightly differing structural analogues of the lead compound may need to be produced to optimise the properties of the drug. This includes minimising side effects. After a series of clinical trials, and billions of dollars of investment, the drug may become available to market.

The work undertaken by medicinal chemists has been instrumental in the improvements observed in the health of society and the increase in life expectancy in modern times. Surgical procedures that are now routine, prior to the development of antibiotics, carried great risk of septicaemia and death. Many diseases caused by harmful pathogens can now be treated. Malfunctions of the body or mind that are understood on a molecular basis can also be cured by medicines. Conditions that were once life-threatening, such as diabetes or heart disease, now have drugs that are available to preserve a healthy life. As people are now living longer, society is faced with the challenges of an aging population. Diseases such as cancer and neurological deterioration in Alzheimer's and dementia are now starting to be understood and drugs, which strive to combat these diseases, are available on the market. It is important to acknowledge how much chemistry has improved our lives.

Finding a Target

2

THE CELL: NATURE'S LABORATORY

Drugs must interact with the body to produce a biological response. On a microscopic level, the tissues which comprise our bodies are made up of individual cells. It is with the cells that drugs perform their function. The cell is a very complex structure and offers numerous targets on which drugs can work. The cell likewise can be considered simply as a reaction vessel; living creatures are composed of chemicals and are obedient to the same laws of chemistry as any laboratory experiment. Knowledge of the principles of chemistry means that medicinal chemists can predict how molecules will interact with a biological target and design drugs that will generate the required response to alleviate the symptoms of a disease. The cell contains thousands of essential molecules, where chemical energy drives their biosynthetic reactions to produce the cell's fundamental components. Small molecules, which are predominantly obtained from the diet, are used to synthesise the giant macromolecules of the cell. Principles of thermodynamics give order to these polymers in this chemical system resulting in them adopting predetermined conformations, dependent on the sequence of their monomer sub-units. These macromolecules assemble to form the vital structures of the cell, such as receptors, transport proteins, enzymes, as well as non-protein structures such as plasma membranes and nucleic acids. It is within this ensemble of molecules that a molecular target for a particular disease must be identified and a drug designed to interact with this target in such a way as to serve as a therapy.

The elements that are prevalent in organic chemistry are indeed the fundamental building blocks of the cell also. Nature carries out its organic reactions within the aqueous environment inside the cell, with water serving as a solvent. Polar molecules are dissolved; those with functional groups containing electronegative atoms that pull the electron density of the molecule towards themselves are held within the aqueous medium, while non-polar hydrophobic compounds remain separate from the internal cellular solution, known as the cytoplasm. This property of different solubility of biological molecules is crucial to the cell, for it governs how molecules interact and react together and how the cell membrane is formed to give the cell its structure and stability. Many of the functional groups commonly used in organic chemistry are also frequently seen in nature: methyl (CH_3), hydroxyl (OH), carboxyl (COOH), and amino (NH_2) recur repeatedly in biology. The small organic molecules found in the cell generally contain up to 30 atoms and have many uses in the cell, such as intermediates for deriving energy from food and units to build polymers which comprise the majority of the cells composition.

The synthesis and breakdown of biopolymers follow a discrete sequence of chemical changes and follow definite rules. As a result, many of the biological compounds that make up the cell are chemically related and can be broadly classified into four major families of small organic molecules: the simple sugars, the fatty acids, the amino acids, and the nucleotides. Sugars are the food molecules of the cell; broken down to create chemical energy. Fatty acids play an important role as the main component of the cell membrane, and amino acids and nucleotides are the sub-units of two crucial groups of biopolymers: proteins and nucleic acids, respectively.

Biological molecules may consist of many thousands, or even millions of these sub-units to produce structures where each atom is precisely linked into a specific spatial arrangement; imperative to determine the macromolecule's properties, which in turn governs their specific function in the cell. The specific sequence by which these sub-units, be it amino acids or nucleotides, are organised carries specific information and generates a biological message that can be "read" through interactions with other molecules. This is how biological macromolecules perform their function: by interacting with the appropriate molecular counterparts, governed by specific intermolecular forces that exist between them, macromolecules can perform roles in the cell such as catalysing chemical transformations, assembly into multi-molecular structures, generate motion, and, most fundamental, storing hereditary information.

The cell is an aqueous environment, where water comprises 70% of the total mass. The remainder is mainly due to macromolecules. These are assembled from their monomer sub-units through particular biochemical mechanisms which specifically control the sequence of the monomers added to the end of the polymer chain as well as determining the appropriate time to terminate the sequence. The macromolecular chain is linked by covalent bonds, which are strong enough to preserve the sequence for a long period of time. This precise sequence dictates the information contained within the macromolecule, but utilization of the information is controlled through non-covalent interactions. These are much weaker bonds which exist between the polar functional groups between different molecules or different parts of the same molecule. These electrostatic interactions govern the three-dimentsional shape of a macromolecule and, therefore, how it will interact with other molecules.

These interactions singularly are too weak to withstand thermal motions; hence, macromolecular structures are held in place by multiple non-covalent interactions. For this to occur, spatial arrangements of the atoms must be precisely matched for a strong interaction. For example, in the case of hydrogen bonding, which is a particular kind of non-covalent interaction between a lone pair of electrons on a sufficiently electronegative atom, namely nitrogen, oxygen, or fluorine, and the electron deficient hydrogen atom bonded to the aforementioned elements. The donated lone pair of electrons must be in the correct orientation relative to the accepting hydrogen atom of the other molecule/other part of the same molecule to satisfy the directionality of the intermolecular force. In biochemistry, usually multiples of these interactions are required between molecules, which leads to the need for interacting molecules to be specific shapes, as in the case of enzyme-substrate complementarity. These exacting requirements account for the specificity of biological recognition.

The number of possible three-dimensional shapes of a macromolecule is restricted by the nature of atoms behaving as hard spheres, with a definite radius meaning no two atoms can overlap, which restricts the number of possible bond angles in a polypeptide chain. This in turn, and in addition to steric factors where atoms hinder one another in a congested region of a molecule, puts constraints on the number of three-dimensional arrangements, or conformations, that the molecule can exhibit. In the case of proteins, steric hindrance from amino acid side chains and the way in which the electrons of the covalent bonds resonate between the amino acids, produce a rigid primary structure, where the amino acid side chains are held pointing outward in particular positions depending on the sequence of amino acids. This feature means that the polymer chain will fold into a predetermined secondary, and ultimately, tertiary structure. Here, nature works to ensure a specific molecule is made, needed for a specific function. If the primary structure had the full flexibility expected for single covalent bonds, the synthesis of proteins would be much more erroneous.

Proteins are built from a repertoire of 20 amino acids; all necessary for protein synthesis, but only 10 of the 20 can be synthesised by humans; the other 10 must be obtained from dietary sources. Protein chains, often thousands of amino acids long, fold into unique three-dimensional structures due to electrostatic interactions within the chain or between other chains to produce the specific structure that governs the function of the protein. Proteins are incredibly important from a biological perspective: they have many functions in the body, attributable to their versatility. Their name, from the Greek *proteios* meaning of first rank named in 1838 by Jons J Berzelius, indicates how crucial these molecule are to the cell because without them, cells would lose their structure, biochemical reactions would not proceed, and cellular communication could not occur. Proteins are classified as either fibrous or globular, depending on this three-dimensional structure. Proteins that are involved in the control of biochemical processes tend to be globular, while protein involved in a mechanical occupation are usually fibrous.

Amino acids are di-functional, containing a basic amine group and acidic carboxyl group. Each of the 20 amino acids is an alpha-amino acid, in other words, the amine substituent is attached to the alpha carbon next to the carbonyl group. Also note that 19 of those amino acids are primary amino acids and only differ in the side chain substituent. Having both acidic and basic groups, amino acids are able to undergo internal acid-base reactions to produce dipolar ions, called zwitterions at a particular pH, known as the isoelectric point, and are amphoteric, acting as both an acid and a base, which is crucial for their involvement in biological catalysis. A polypeptide is synthesised from amino acids by the condensation reaction between the carboxylic acid group of one amino acid with the amino group of the next to form an amide linkage, or peptide bond, via an enzyme-controlled pathway.

Proteins have three-dimensional folded structures; the shape of which is very specific, depending on the almost infinite variety of amino acid sequences, and is predetermining for their given function. Proteins can fold into a regularly repeating secondary structure; either an α-helix or a β-pleated sheet. These structures are governed by the primary structure; as hydrogen bonds are formed between different amino acids in the chain, or covalent bonds in disulphide bridges between cysteines. In an α-helix, the primary chain twists into a coil about a central axis, while an undulating sheet-like structure is obtained in the formation of a β-sheet: which of these different structures is formed depends on the precise hydrogen bonding that arises from the given sequence of amino acids. Further folding may then occur to produce a tertiary structure. For proteins composed of several constituent polypeptide chains, the final folded structure is referred to as the quaternary structure. These unique shapes have discrete implications regarding the function of the protein.

The proteins that bind together to produce a quaternary structure, which is the finished functional protein, are organised in a particular arrangement to produce a structure with the correct properties to perform a specific role in the cell. For example, haemoglobin carried in red blood cells consists of four sub-unit peptides, each with a cofactor, called the haem group, with an iron atom centre to which an oxygen molecule can bind. Each haemoglobin molecule can therefore carry four molecules of oxygen and the precise arrangement of the four polypeptide sub-units is crucial for the conformational changes that the protein undergoes to facilitate oxygen binding.

Half the dry mass of a cell is made up of proteins, which are responsible for the structural integrity of the cell as well as catalysis and molecular recognition. Proteins are of crucial importance in the body; having many key functions: enzymatic catalysis, transport and storage of small molecules, coordinated motion (make up muscles), mechanical support (collagen, a fibrous protein, in skin and bone), immune protection (antibodies are highly specific proteins), generation and transmission of nerve impulses, and control of growth and differentiation of cells (involved in DNA replication). The information for making these proteins is stored as genes on DNA.

The instructions for producing proteins are encoded on genes, sections of DNA, which are transcribed on to RNA before being translated into a protein chain at ribosomes; cell organelles designed for this operation. This is a very intricate enzyme-controlled biological pathway, with a high degree of complexity. Proteins produced from this process then adopt their specific three-dimensional shape and move to their intended location to perform a given function in the body. Nucleotide triplets code for particular amino acids; the order in which these triplets are assembled in the gene determines the sequence of amino acids in the protein and therefore ultimately governs the proteins shape and function. The different side chains of each of the 20 amino acids can form a variety of hydrogen bonds; hence the observed variety of possible structures. Proteins are very versatile molecules that can have many functions, governed by their structural organisation. For globular proteins such as those involved in catalysis and molecular recognition, the final folded three-dimensional shape of a polypeptide, the tertiary structure, may constitute one protein domain, and several globular units comprise the functional quaternary structure, as with haemoglobin. In order for polypeptides to assemble into a functioning quaternary structure, precise interactions between these molecules must be in place.

Intermolecular interactions are of fundamental importance in biochemistry. The forces that exist between molecules are responsible for the assembly of protein sub-units to form the quaternary structure. Intermolecular forces are crucial for biological catalysis with enzymes, molecular recognition via substrate-receptor binding, and DNA processes as well as many other biochemical processes. In order to bind, the molecules must come into contact to enable formation of the non-covalent interactions. Molecules move randomly due to thermal motion as they collide and bounce off each other. These collisions must have sufficiently low energy to permit a binding interaction to occur between molecules, in other words, collide with less energy than the enthalpy of bond formation for the intermolecular force. As such, the rate of complex formation between two molecules is said to be diffusion-limited. Diffusion is slower for large molecules with a larger collision cross-section than it is for small molecules. Furthermore, the rate of complex formation is impeded by the need for the binding surfaces to be orientated so that the interacting surfaces can fit together. This is associated with a reorganisation energy, hence most collisions do not result in bond formation. These dynamics need to be considered in the development of pharmaceuticals to assess the potency of a drug. This is referred to as pharmacokinetics. Also, understanding the thermodynamics of biochemical processes, such as protein-DNA interactions, substrate-receptor binding, and enzyme-substrate complex formation is a necessity for designing an effective drug to work on these kinds of targets. The strength of the binding is measured with an equilibrium constant K = [HS]/[H][S], which gives the ratio of the concentration of the complex in solution over the concentrations of the separate host and substrate. In other words, the larger the equilibrium constant, the more host-substrate complex there is in solution, the stronger the binding. The strength of binding has important implications for drug design. For example a drug working on an enzyme host will have a potency that depends on the strength of the binding, which can be modified through understanding of the thermodynamics of molecular interactions.

The rate at which chemical reactions proceed in the cell is exceedingly fast. Many cellular reactions are catalysed by enzymes, which are needed as many reactions would otherwise proceed at imperceptible rates at body temperature. They can catalyse biological reactions on the order 10^3–10^6 reactions per second. The interactions between enzymes and substrates are governed by thermodynamics and are diffusion-limited, but the rate of reaction is enhanced by providing an alternative route for the reaction to proceed with a lower activation energy. This is achieved by binding of the enzyme, which can distort the substrate molecule into a more reactive conformation, and the kinetic aspect is improved by holding one of the reacting molecules, the substrate, in place, in the correct orientation for reaction with a second molecule.

For proteins that function as a catalyst, namely enzymes, the conformation of the protein governs its chemistry. The precise folded structure gives a unique surface to the protein, where neighbouring residues interact in such a way as to alter the chemical reactivity of selected amino acid side chains. Particular residues may be orientated in such a way as to optimise binding interactions with a substrate, forming complementary hydrogen bonds. Furthermore, neighbouring parts of the polypeptide chain may interact in a way to exclude water molecules from the active site, which would compete with the substrate for hydrogen bonds. While it may be difficult to conceive that a structure can restrict access to a small molecule like water, it is often energetically unfavourable for water molecules to separate from the hydrogen bond network of the aqueous medium to habituate a crevice on the protein surface. The clustering of amino acid side chains due to the particular folded structure can activate normally unreactive side chains. Negatively charged side chains can be forced together against their mutual repulsion by the tight folded structure and create an attractive binding site for positively charged substrates. When the necessary reactivity cannot be achieved by simply having the correct organisation of side chains, proteins can utilize other non-polypeptide molecules called cofactors. These bind to the protein surface and serve to exploit the reactivity of these molecules that they acquire when bound to the protein; an example being the iron in haem and cytochrome C in a protein-bound metal ion complex with complicated organic chelates.

The precise shape of proteins means that enzymes catalyse specific chemical reactions. This is referred to as enzyme-substrate specificity, as only certain molecules are able to bind to a given active site. This selectivity is crucial for reactivity, as enzymes speed reactions by selectively stabilizing transition states. A reacting molecule adopts a high energy conformation (the transition state) and precise molecular interactions are required from the enzyme to stabilise this structure so that the reaction can proceed correctly. The catalytic ability of enzymes is far superior to any synthetic catalyst, and this efficiency is due to several factors. As previously mentioned, the enzyme increases the local concentration of the substrate at the catalytic site by holding it in the active site in a position with the correct orientation for the reaction. Most important are the binding energy contributions. Upon substrate binding, the enzyme modifies the geometry of a substrate molecule to produce an unstable structure with a different electron distribution and a much higher free energy value. This structure, the transition state, is therefore higher in energy than the substrate; being in a reactive conformation. The activation energy, the minimum amount of energy needed for the reaction to proceed, is equal to the difference in energy between the original substrate and the TS, so energy input is needed for the reaction to proceed. The reason these reactions can occur much more frequently than they otherwise would in the absence of a catalyst is because the binding energy of the enzyme lowers the energy of the transition state, bringing it closer in energy to the favoured conformation of the original substrate, hence lowers activation energy, and therefore less energy input is needed. Without enzyme intervention, much more energy would be needed to distort the substrate to the reactive conformation therefore the reaction would be energetically less favourable and proceed at a slower rate.

The way in which enzymes promote a reaction, by breaking and making covalent bonds, can proceed via different mechanisms. In the absence of enzymes in a laboratory scenario, peptides can be hydrolysed using either an acid or a base catalyst. The precise positioning of amino acid side chains in the active site means that enzymes can have the unique ability to utilise amphoteric residues for acid an base catalysis simultaneously because the acidic and basic residues are prevented from combining with each other, as they would do in solution, due to being bound to the rigid protein framework.

As a result, a specific substrate, with the precise requirements to fit the active site, will be positioned in such a way that the acidic and basic residues are in the correct position to disperse the electron density of the substrate so that electron density is withdrawn from the substrate by the acidic residue which increases susceptibility to the appropriate nucleophile, for example water in the case of hydrolysis. Meanwhile, water is made into a stronger nucleophile by the action of the basic residue pushing electron density onto the reactive centre, in this case oxygen, resulting in the rate of reaction being increased.

The reaction profile in Figure 2.1 illustrates how the energy difference between the substrate and the transition state at the top of the curve, which is the activation energy, is lowered in formation of an enzyme-substrate complex. Less energy is required for the reaction to proceed, so more encounters between the enzyme and the substrate will lead to a successful reaction, hence rate is increased. The mechanism for hydrolysis of a peptide bond explains how the catalytic process operates. The negative charge of the carboxylate pushes electron density onto the electronegative oxygen atom of the water molecule, making it a strong nucleophile. Meanwhile, the electronegative nitrogen atom of a second amino acid residue pulls electron density from the C=O bond, creating a partial positive charge on the carbon, making it very susceptible to nucleophilic attack from the oxygen lone pair of electrons, hence the reaction proceeds quickly.

The example given above shows how proteins can be catalytically hydrolysed, for example during digestion of food. Enzymes may also increase rates of reactions in other ways. For example, they may form a covalent bond with the substrate, which therefore becomes attached to a residue or coenzyme in the active site. This can have the effect of weakening the covalent structure of the substrate, which will

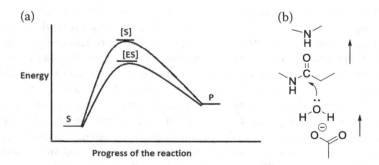

FIGURE 2.1 A Schematic Reaction Profile Is Shown in Diagram (a) example for Hydrolysis of a Peptide. Diagram (b) Shows the Mechanism for Hydrolysis of a Peptide Bond.

then react more readily with a second molecule, which causes cleavage of the bond to the enzyme, which is now free to catalyse another reaction. This is how serine proteases work, for example.

The diversity of proteins required for a cell to function are all transcribed from the cells instruction manual: deoxyribonucleic acid (DNA). The DNA resides within the cell nucleus, which is effectively the control centre for the cell. The nucleus is a membrane-bound structure containing pores to permit the selective passage of certain molecules required for cellular activity at this organelle. Nucleotides and polynucleotides, such as ribonucleic acid (RNA), continually diffuse in and out of the nucleus. The code for building the proteins of the cell are carried from the control centre on RNA to the cytoplasm where the information contained on these molecules is interpreted and translated into an amino acid sequence to assemble a protein to be used in the cell. Protein synthesis, along with most other cellular activities, is controlled from the genes contained within DNA.

The DNA molecule is a long unbranched polymer composed of nucleotide sub-units, which contain the bases: adenine (A), cytosine (C), guanine (G), and thymine (T). The nucleotides are linked together covalently with phosphodiester bonds between the 5' carbon of one deoxyribose group and the 3' carbon of the next. These linkages form the sugar-phosphate backbone of the DNA molecule to which the four bases are attached. Two strands of DNA combine via intermolecular bonding between pairs of bases; one from each strand to form the complete molecule. These base pairs are instrumental in determining the characteristic structure of DNA. Two DNA chains combine through hydrogen bonds between the bases like runs of a ladder and cause the structure to twist into the α-helix shape. This structure was discovered by Watson and Crick in the early 1950s from x-ray diffraction studies. Their model of the DNA structure revealed that the DNA bases were inside the double helix and that the bases of one strand were extremely close to those of the other, which results in the need for specific base pairing between a large purine base (A or G, which contain a large double ring structure) and a smaller pyrimidine base (T or C, each of which has a single ring) on the other chain. These requirements for spatial arrangements of atoms to satisfy hydrogen bonding results in complementary base pairing between A and T and G and C, where two and three hydrogen bonds are formed, respectively, between the bases. Alternative base pairing would result in either too greater separation for hydrogen bonding, or steric congestion between pairs large bases.

Different sequences of these base pairs comprise a gene. Genes are the information containing elements of DNA that code for a particular protein and produce hereditary characteristics, such as eye colour. In a population, there may be different forms of the same gene. In the example of eye colour, the gene coding for the pigment in the iris varies between people who have different coloured eyes. The different forms of a gene are called alleles. Remarkable diversity of proteins is achieved from these four bases. The unique sequence of the bases on a gene governs the specific sequence of amino acids during protein synthesis. The DNA sequence exhibits a triplet code, where a particular combination of three base pairs codes for a particular amino acid when translated from the RNA sequence at the ribosome, which was transcribed from DNA. A particular sequence of triplets codes for a specific protein.

During cell division, identical copies of these genes must be transmitted to each daughter cell. The information contained within the genes is written as the unique linear sequence of nucleotides that make up that section of the DNA. A consequence of complementary base pairing means that each strand contains identical information and can serve as a template for the replication of that information. This is the principle that underpins DNA replication and protein synthesis. Errors may occur during DNA replication and this causes mutations.

All the cellular processes that proceed in a controlled environment within the cytoplasm are in isolation from the disruptive influences of the external environment. This is attributable to the protective qualities of the cells' outer membrane. The plasma membrane is crucial to the cell because it maintains the internal environment within the cytoplasm and separates the operating systems of the cell from the extracellular environment, which is essential for the sensitive biochemical processes that are the habitual conduct of the cell. The plasma membrane enables selective passage of substances into and out of the cell to sustain the delicate chemical system within. Structures inside the cell, called organelles, are also comprised of a plasma membrane. Among these membrane-bound organelles are the Golgi apparatus, endoplasmic reticulum, mitochondria, as well as the cell nucleus; all of which play a crucial role in sustaining the dynamic operating systems inside the cell. The contents within these organelles, related to their specific function, are characteristically different to the rest of the cytoplasm; maintained by the plasma membrane that keeps these systems separated.

Organelles perform different roles in the cell. The mitochondria are responsible for energy generation. Through the process of respiration, they manufacture adenosine triphosphate (ATP); a source of chemical energy. Modification of selected molecules is undertaken at the Golgi apparatus and/or the endoplasmic reticulum. Molecules manufactured by the cell, or assimilated from outside, may need to be changed to make them suitable for a given biochemical process. The Golgi apparatus and endoplasmic reticulum are the organelles with the capacity to perform this role. The nucleus, as previously mentioned, is where DNA is contained and is the control centre for the cell. Having a porous membrane enables passage of selected molecules to facilitate communication between the nucleus and the rest of the cell to ensure that instructions are received for the essential biochemical process of the cell to continue functioning without error.

The activities of these organelles and differentiated cells are governed by the specialised membranes. They have unique transport proteins for selective transport of molecules or ions, related to their function. Neurones, example rely on particular transport proteins used to conduct nerve impulses by generating a potential difference (voltage) across the cell membrane by transporting sodium and potassium ions, which gives a difference in charge between the inside of the cell and the tissue fluid outside. Mitochondria undertake aerobic respiration through which chemical energy in the form of ATP is produced through biochemical processes involving the transport of hydrogen ions across specialised membranes. Release of enzymes, other protein moieties, hormones etc. tend to be assisted by transport proteins. The release of insulin from cells of the pancreas to control blood sugar is a notable example because errors that can arise in receptor signalling or transport of this small protein-based hormone lead to the disease diabetes. The cell membranes contain specific receptors needed for a specific response related to cells function in the tissue, such as receptors for insulin to control blood sugar level. Receptor proteins transfer information across the cell membrane rather than ions/molecules. The cell is then able to instigate a response to this information, such as release hormone, neurotransmitter, initiate cell division—whatever response is required. Confusion in cell signalling can therefore obviously lead to problems.

Although the biological membranes have different functions, the general structure of the plasma membranes are similar. It is composed of a phospholipid bilayer in which the characteristic proteins are embedded. This consists of a very thin film which encloses the cell or organelle and is described as a "fluid mosaic" as the lipid and protein molecules are free to move around in the plane of the membrane because the structure is held together by a dynamic array of non-covalent interactions that continually break and re-form, permitting the motion of molecules across the membranes' horizon.

The phospholipid bilayer establishes due to the oil-like nature of the lipid. When an oil is dropped into water, it will form a layer at the surface (provided it is less dense than water) separated by a phase boundary, which is the interface between the hydrophobic non-polar hydrocarbon chains of the oil molecules and the polar water molecules that are repelled by the oil molecules. As droplets of oil within the volume of the water ascend, they too are repelled by the water molecules and form a globule, or a hollow vesicle, due to the hydrophobic interaction. This interaction between lipid and water to form a vesicle is how a cell membrane is derived. The difference is that the cell membrane is composed of a bilayer. This originates from the nature of the phospholipids that comprise the cell membrane. They consist of a polar head, the phosphate, which interacts with the water molecules at the interface, and a hydrophobic hydrocarbon tail, which is repelled by the water, so is buried internally within the membrane. Here, the hydrophobic tails interact with mutual attraction to hold the structure together. This results in a double layer about 5 nm thick that is relatively impermeable to water-soluble molecules, so allows for selective transport of substances in and out of the cell to maintain a constant internal environment; in other words an isolated chemical system.

The composition of the phospholipid bilayer affects the fluidity of the structure. Eukaryotic cell membranes contain especially large amounts of cholesterol, which bind to the polar head group of the phospholipid, thus reducing the mobility. Regions of the cell membrane containing large amounts of cholesterol are therefore less deformable. The higher cholesterol composition also has the effect of increasing the permeability barrier. The phospholipid bilayer is less permeable in this instance; predominantly water and lipid-soluble small molecules are the only substances that have ease of passage. This is important for the consideration of drug design if the target is inside the cell.

Most of the specific functions carried out by the plasma membrane are conducted by membrane proteins. As such, the type and quantities of these proteins in plasma membranes varies a great deal, depending on the function of that membrane. In myelin membranes, which function as the electrical insulators in nerve cells, less than 25% of the membrane mass is protein. By contrast, in the inner membranes of mitochondria, which are involved in energy transduction, are about 75% protein, by mass. Typically, cell membranes will have approximately 50% of their mass being proteins. Proteins molecules are much larger than the lipid molecules that comprise the bilayer, so there are many more lipid molecules in the membrane than there are proteins. The membrane proteins on the cell exterior will often have oligosaccharide (carbohydrate) molecules attached to them, which form a coat on the cell surface.

There are different ways in which membrane proteins can be associated with the phospholipid bilayer. Transmembrane proteins extend through the lipid bilayer and, like the lipids, are amphipathic, consisting of hydrophobic regions which interact with the lipid tails within the membrane, and hydrophilic regions that protrude the membrane and are exposed to water. This morphology is determined by the position and nature of the amino acid side chains. The final tertiary or quaternary structure of the protein will place side chains with polar functional groups to the outside in the regions exposed to water, while greasy side chains that are hydrophobic interact with the lipid membrane.

Other membrane proteins are located within the cytoplasm; attached to the cell membrane by means of covalent bonding with fatty acid chains, while other membrane proteins are entirely exposed to the extracellular medium and are attached to the cell membrane by covalent bonds to specific oligosaccharides. Additional proteins may be bound to these integral membrane proteins by non-covalent interactions and are known as peripheral membrane proteins. These types of proteins have different functions. The transmembrane proteins are often involved in molecular transport across cell membranes. Extrinsic proteins serve as cell signalling receptors.

Transport across plasma membranes is a crucial part of a cells existence. The cell membrane presents a barrier to most polar molecules, which is important for maintaining concentrations of solutes in the cytoplasm. Likewise, the membrane-bound organelles within the cell can have a specific concentration of molecules contained within; different from that of the cytoplasm or extracellular medium. However, critical substances required by the cell must have a means of entering the cell as well as the removal of waste products. This is where the key role of transmembrane transport protein comes into fruition; as they are responsible for transporting these water-soluble molecules across the plasma membrane. A given transport protein will be responsible for assisting the movement of closely related groups of organic molecule, or a specific ion, across the membrane. There are two classes of membrane transport protein: carrier proteins and channel proteins. Carrier proteins have moving parts, activated by the chemical energy source ATP, that mechanically move small molecule across the membrane. This is known as active transport. Channel proteins form a narrow hydrophilic pore that enables the passive movement of inorganic ions, known as facilitated diffusion. By these mechanisms, the cell can create large differences in composition between the internal environment and extracellular medium. This is essential for specialised cells to perform their role in the body.

To elaborate by revisiting previous examples: the pancreatic cells must release insulin into the blood stream to control blood sugar level; in the mitochondria, hydrogen ion transport is required for ATP synthesis; and neurones must transport sodium and potassium ions to produce an electro-chemical gradient. Transport proteins are instrumental for all these processes. One can acknowledge that malfunctions in these processes are likely to result in disorders. Cystic fibrosis is a disease caused by a defect in a transport protein for chloride ions. Diabetes from malfunctions in the control mechanisms of homeostasis; Alzheimer's, dementia, and other neurological disorders are all challenging diseases at this time and understanding the molecular processes behind the conditions will help in the development of treatments.

For a cell to operate normally as a single unit, besides the requirement of transport proteins to allow passage of essential nutrients, intracellular signalling proteins are needed to generate a response to changing internal conditions. In multicellular organisms such as ourselves, cells must also interact with one another. An array of cell surface receptors enable response to many chemical massagers simultaneously and sensitively. Extracellular signalling molecules are recognised by specific receptors on the surface membrane, or within the target cells. There are hundreds of kinds of signalling molecules with which eukaryotic cells communicate, including proteins, small peptides, amino acids, nucleotides, fatty acid derivatives, steroids, retinoids and even dissolved gasses, such as nitric oxide and carbon monoxide. These communicating agents may be released from the cell by exocytosis, where they are carried in a lipid vesicle, or by simple diffusion through the plasma membrane. Some may remain bound to the cell if their purpose is signalling to local cells. The recipient cells, to which the communication is directed, will have specific receptors that are a complementary fit with the signal molecule and upon binding propagate a cellular response.

There are three forms of signalling mediated by secreted molecules: paracrine, synaptic, and endocrine. Signalling molecules secreted by cells to act as local mediators, which only effect cells in the immediate environment, must not be allowed to diffuse too far, so are rapidly taken up by the neighbouring cells, or destroyed by extracellular enzymes, or immobilised by the extracellular matrix. This is called paracrine signalling. For multicellular organisms to be able to coordinate cell behaviour across the entire organism, some signalling molecules must travel far afield to distant cells. This is achieved in two ways: by networks of nerve cells and by the action of hormones. Synaptic signalling involves routes of neurones along which electrochemical impulses travel to stimulate the release of chemical signals called neurotransmitter, which carry the signal on between neurones across gaps called synaptic junctions and propagate the electrochemical impulse in the adjoining neurone. Endocrine cells release hormones, which are signalling molecules that travel in the bloodstream of an animal (or sap in plants) and thus distribute widely throughout the body, enabling signals to be carried to distant cells. Since this process relies on diffusion, it is much slower than synaptic signalling.

Autocrine signalling involves cells secreting signalling molecules that can bind to its own receptors and coordinate decisions between groups of identical cells. This generates the "community effect" observed between cells and enhances the effect of signalling when groups of cells carry out the response simultaneously. Activity can also be coordinated between cells through gap junctions, which can form between cells with closely apposed plasma membranes, where the cytoplasm of neighbouring cells are joined directly by narrow water-filled channels. This enables the exchange of small intracellular signalling molecules, such as Ca^{2+} and cyclic AMP, therefore gap junctions allow cells to communicate directly unrestricted by the barrier presented by intervening plasma membranes.

Cell signalling processes allow the cells that make up multicellular organism to communicate with one another; be it their neighbours or distant cells elsewhere in the body. Any given cells may be exposed to hundreds of signals, not all of them relevant to its particular function. In order to perform their specific role in the body, each cell is programmed to respond to specific combinations of signalling molecules. Furthermore, different cells respond differently to the same chemical signals. This means that the behaviour of cells can be controlled in highly specific ways. These signalling molecules generate a response by binding to cell surface receptor proteins, which act as signal transducers, where the binding of the signalling ligand induces an intracellular response that alters the behaviour of the target cell. There are three known classes of cell surface receptor, defined by the mechanism of transduction used. Ion-channel-linked receptors are those involved in the rapid synaptic signalling between neurones during an action potential (nerve impulse), mediated by neurotransmitters. G-protein-linked receptors indirectly regulate the activities of target proteins bound separately to the plasma membrane, for example enzymes or channel proteins. The mediator for this process is called "G protein", which binds to the receptor, and then upon binding of the signalling molecule to the receptor, becomes activated, and then departs to bind to, and activate, the target enzyme or channel protein. The third type of receptor are enzyme-linked receptors, which become activated by a signalling molecule and prompted into acting as an enzyme.

Chemical messages are critical for coordinating cellular activities and maintaining normal function of the body. They are responsible stimulating the body's organs to operate, prompt the growth and repair of tissues, and maintain homeostasis. In order to carry out these processes, cells must manufacture the appropriate raw materials, such as proteins and nucleic acids, which are needed for DNA replication and the enzymes required for cell division. Malfunctions at any stage of these sensitive biochemical processes can lead to disease. Understanding the molecular basis of diseases aids medicinal chemists in designing molecules that have the required properties needed to target the erroneous receptor, protein, gene; whatever may be the source of the disease and interact with the target in a beneficial way to mitigate the symptoms of the disease and serve as a therapy. Medicines are made by finding a target and designing a drug that will influence the cellular activity associated with the target in a way that can restore good health.

TARGETING THE CELL

Drugs operate in the cell by interrupting a biochemical process such that there is a change in the activity of the cell, causing a noticeable effect. They may enter the cell via transport proteins in the cell membrane, or if they are suitably hydrophobic pass straight through the phospholipid bilayer. On the other hand, the drug might target receptors on the outer surface of the cell. How a drug reaches the target is an important factor of consideration for the medicinal chemist. The drug molecules must reach the site of action in sufficient concentration to be effective and this may affect the route of administration of the compound. Compounds that are not assimilated well when taken orally may have to be injected. Chemists can alter the properties of the molecule to make it more orally bioavailable, but this must not impede the interaction of the drug with the target. Drugs can be designed to target specific enzymes, cell receptors, and even nucleic acids. A drug molecule's ability to interact with the target is paramount to determining its efficacy.

Drugs that target enzymes are designed to inhibit their normal operation. This can be achieved in different ways. Competitive inhibitors mimic the molecular structure of normal substrate so that the drug can bind with a complimentary fit to the active site of the enzyme. This has the effect of blocking the active site, preventing entry of the normal substrate. The necessary reaction with the normal substrate cannot proceed, hence the biological process is subdued. The extent of this effect depends on the concentration of the drug, which in turn determines how many active sites are inhibited out of the plethora of catalytically available enzymes. Also important is the strength of binding of the drug to the active site, which effects the length of time that the drug remains in the active site; impacting the probability of normal biological catalysis happening. The nature of non-covalent interactions; being changeably broken and re-formed results in inhibition occurring dynamically. The weaker the intermolecular forces between the drug and the active site, the greater the proportion of unencumbered enzymes at any one time and the biological process will be inhibited to a lesser extent. The drug must be designed to optimise non-covalent interactions in order to be effective. Alternatively, the drug molecule could perhaps be designed to undergo reaction once in the active site to form a covalent bond to an amino acid residue. This is an irreversible form of inhibition and renders that enzyme molecule redundant.

Not all enzyme inhibitors operate within the active site. Another method is non-competitive inhibition, where drugs can be designed that bind to a non-functional part of the enzyme surface and this physical binding may cause distortion of the protein structure; having the effect of changing the shape of the active site so that the substrate will no longer undergo complementary fit with the enzyme. Hence, the normal biological function is interrupted.

Enzyme inhibitors have been used widely in medicine. To combat infections from microorganisms; shutting down enzymes that are crucial to the function of the bacterial cell will kill the cell or prevent proliferation. It is possible to selectively target bacterial enzymes without effecting our own due to the large biochemical differences between bacteria and ourselves. For example, some of the first antibiotics were the sulphonamides. These acted as competitive inhibitors and medicinal chemists synthesised a library of these compounds to optimise binding interactions to improve efficacy. These drugs were the antibiotics of choice prior to been superseded by penicillin, which also function as competitive enzyme inhibitors, except on a different target. In the fight against viruses, successful antiviral drugs have been developed that work against viral enzymes. Acyclovir for the treatment of herpes and saquinavir for HIV are both enzyme inhibitors. Besides battling against foreign invaders, enzyme inhibitors can be utilized to work against the body's own enzymes and thus regulate and offer control over the cells biological operations. Anticholinesterases are an example where inhibitors of enzymes were developed for control of problems with the nervous system.

Inhibition of enzymes interrupts cell biochemistry and subdues the biological process for which that enzyme is responsible. Many of these biological processes are activated by cell signalling processes. Operations of the cell can therefore be interrupted from the start by targeting the receptors for these chemical messages.

In order to maximise intermolecular binding interactions, the receptor protein changes conformation to accommodate the chemical message. This change of shape in the receptor induces a cellular response to the signal. It is therefore clear that to design a drug that targets a specific receptor, it must satisfy these non-covalent interactions and partake in complementary binding. Essentially, the drug must mimic the natural substrate in an analogous way to enzyme-substrate complex formation. Moreover, given that cellular receptors control biochemical processes within the cell, drugs can be designed that either enhance or suppress cellular activities. Drugs that are designed to closely mimic the natural substrate are agonists; they have the effect of increasing the concentration of chemical messages so that the receptor is more frequently activated and the biochemical process to which that receptor is linked will be enhanced. In some instances, a drug may be required to supress the activity of the cell: antagonists are designed to mimic the natural substrate closely enough to interact with the receptor, but are sufficiently different as to not trigger a response upon binding. This has the effect of decreasing the availability of receptors and suppress the activity.

To design an agonist, the drug must have the correct binding groups placed in the right positions for binding, and indeed be the correct size for the binding site of the receptor. The binding groups are those that form noncovalent bonds with the receptor and may comprise polar functional groups, producing electrostatic interactions, or greasy hydrocarbon sections that will form hydrophobic interactions. When designing a drug, these groups will be placed at precise positions of the target molecule to maximise binding forces with the receptor. To achieve this, a medicinal chemist must have an understanding of how the messaging molecules interact with the receptor. This is known as pharmacodynamics. Typically, this can be done using x-ray crystal diffraction studies of the isolated target protein to elucidate the nature of the protein's binding site. From this information, binding interactions with the drug can be inferred and tailored to control activity of the drug inside the organism. This is referred to as pharmacokinetics.

When a drug is required to supress the activity of a receptor, binding interactions must be designed so that the drug behaves as an antagonist. Molecular modelling and x-ray crystallography can be used to reveal the structure of the binding site, which enables deduction of drug pharmacodynamics so that binding groups can be positioned on the target molecule to produce a drug that binds to the receptor, but in such a way that the change in conformation of the protein receptor is incorrect for propagating a response to the signal. Generating a library of target molecules, each with slightly different binding properties to find a potentially successful drug; the suite of molecules can be screened against the binding site to optimise the pharmacokinetics and yield a final medicinally useful compound.

There are also alternative ways that antagonists can be designed, without complementary binding to the receptor site. In an analogous method to non-competitive enzyme inhibition, allosteric antagonists may bind to a completely different part of the receptor protein, resulting in distortion of the receptor site, which is therefore unable to accommodate the natural substrate. Alternatively, binding can take place with amino acid residues outside of the receptor site and the steric bulk of the drug molecule may block entry to the receptor binding site and prevent access of the natural substrate. This is referred to as the "umbrella effect". It is a useful strategy because it may be difficult to develop a drug to target the receptor binding site if molecular modelling of this region is unclear.

Described so far, drugs can be designed to interact with enzymes and protein receptors to control specific biochemical processes and have a medicinally useful effect. The habitual operations of the cell are controlled ultimately by the genetic code inscribed on the cell's DNA, which resides within the nucleus. Drugs can be designed to target DNA, which has huge potential for medicinal therapies. Some of the most notable drugs that target DNA are used in cancer treatment. There are different ways that these drugs act on the DNA molecule and are classified accordingly, including intercalating agents, alkylating agents, and chain cutters. In therapies for genetic illnesses, which arise from abnormalities in the patient's DNA, molecular biology and genetic engineering has produced rapid advances in the understanding of genetic diseases, such as haemophilia, cystic fibrosis, and many others. Many examples of drugs designed to target enzymes are involved in the battle against pathogens: microorganisms that cause infectious diseases. This has, in the past, proved a successful strategy to combat infectious diseases.

Antibiotics: The Need for Innovation

3

A HISTORICAL PERSPECTIVE

In the past, bacterial infections had a high likelihood of resulting in morbidity; as the ailed person relied on their body's immune system to overcome the infection. Very few remedies were available that could alleviate the infection; these mostly included medicinal herbs and dubious tonics, which have been used throughout history. Prior to the development of antibiotics, infection from pathogenic bacteria was a major cause of death. The development of antibiotics in the 20th century led to a dramatic increase in life expectancy. From a cultural point of view, these drugs are perhaps among the most valuable developments in 20th century medicine.

Bacterial infections continue to be an important health issue, particularly in developing countries, and as resistant strains develop, medicinal chemists must continue to develop new strategies to 'battle the bugs'. Tuberculosis is one of the top ten causes of death worldwide, with ten million people falling ill with TB in 2017, and 1.6 million died from the disease, as reported by the World Health Organisation (WHO). However, the incidences of TB globally are falling by 2% each year, and thanks to effective diagnosis and treatment, an estimated 54 million lives were saved between 2000 and 2017. However, multi-drug resistant TB (MDR-TB) remains a public health crisis and a risk to security, emphasising the necessity for new front-line drugs to be developed. The outbreak of antibiotic-resistant pathogens are an ever-present challenge to medicine, and novel compounds must continually be developed to tackle these germs. Infections that cannot be readily treated can lead to sepsis, which is the clinical manifestation of acquired infections, which arises when the body's response to the infection injures its own tissues and organs. It is thought to affect tens of millions of people, potentially resulting in millions of deaths, according to studies reviewed by the WHO. Many diseases caused by bacteria are prominent in developing countries, where medical infrastructure is limited. For example, cholera outbreaks result in the loss of tens of thousands of lives, even though many cases can be treated effectively with oral rehydration therapy (ORS), if not antibiotics. Pneumonia accounts for 16% of deaths of children. Pneumonia caused by bacteria can be treated with antibiotics, but only one-third of children who fall ill of the disease receive the antibiotics they need. *Staphylococcus* infections are a relevant problem, even in developed countries, particularly with the upsurge of MRSA. This reflects the need for new drug development.

Historical uses of antibacterial herbs or potions has been documented for several countries, including the Chinese using mouldy soybean curd to treat boils and wounds. Greek physicians used wine for treating wounds; myrrh as well as inorganic salts and honey was used on wounds in the middle ages. Of course, these people did not understand that infections were caused by microorganisms. Indeed, bacteria were only first identified in the 1670s by Leeuwenhoek while using microscopes to observe microscopic entities in water; it was not until the 19th century that their link with disease was realised, thanks to experiments by Pasteur.

Edinburgh surgeon Josef Lister was an early advocate of a 'germ theory of disease' and used carbolic acid as an antiseptic to sterilise the operating theatre to avoid infecting his patients. This improved surgical survival rates drastically. Later in the 19th century, scientists such as Koch were able to identify the microorganisms responsible for a particular disease. Paul Ehrlich was a pioneer in finding chemical agents that could interfere with the proliferation of microorganisms, while using concentrations that were tolerable by the patient; the so-called magic bullet strategy. This toxicity can be represented by a therapeutic index. By 1910, Ehrlich had successfully developed the first fully synthetic antimicrobial drug. This was the arsenic-containing compound Salvarsan, which is particularly noted for the treatment of syphilis and was used until 1945 when it was replaced by penicillin.

However, Salvarsan was only effective against a small number of bacteria; a much broader antibiotic was required. In 1935, it was found that a red dye called prontosil was effective against streptococcal infections in the blood and this lead to the development of a family of broad spectrum antibiotics known as the sulfa drugs, or sulphonamides, which were the only effective treatment against systemic infections until penicillin became available in the 1940s.

Despite been discovered in 1928, it was not until 1941 that effective methods of isolating penicillin were developed by Florey and Chain. This drug revolutionised the battle against infection. However, penicillin is not effective against all types of infections. Since it was discovered that penicillin is a toxic fungal metabolite that kills bacteria and allows the fungi to compete for nutrients, it encouraged scientists to investigate microbial cultures from across the globe in search of other possible therapeutic agents. In 1944, the systematic search of soil microbes revealed the antibiotic streptomycin, which extended the range of therapy to the tubercle bacillus and a variety of gram-negative bacteria. Continued research led to the discovery of the other major classes of antibiotic: peptide antibiotics, tetra-cycline antibiotics, macrolide, and cyclic peptide; and synthetic agents including cephalosporin C, isoniazid, nalidixic acide, ciprofloxacin, and many others.

Thanks to antibiotics, a wide range of diseases have been brought under control, including syphilis, tuberculosis, typhoid, bubonic plague, leprosy, diphtheria, gas gangrene, tetanus, and gonorrhoea. Antibiotics represent a great achievement for medicinal chemistry when considering the hazards that society faced in the period before penicillin. Within living memory, mothers risked septicaemia in child birth; ear infections in children leading to deafness were common; pneumonia was a frequent cause of death in hospitals; tuberculosis was a major problem; minor injuries could lead to severe infection, requiring amputation of a limb; and the risk of peritonitis lowered the success rate of surgical operations. Perhaps we take antibiotics for granted today; they are readily available and used frequently for mild conditions such as a chesty cough, and few of us appreciate the profound effect that these drugs have had on quality of life.

The success of antibiotics relies on the drugs being damaging to bacteria and acting selectively against the bacterial cell wall and not affecting cells of the patient. This can be achieved through understanding of the bacterial cell, as they tend to have different structures and biosynthetic pathways. Bacterial cells are classified as prokaryotic and have many differences to eukaryotic animal cells. Bacterial cells do not have a defined nucleus to contain their DNA like animal cells do. Instead, the DNA is held loosely in the cell, or on a ring structure, called a plasmid. Animal cells have many other organelles besides the nucleus, such as endoplasmic reticulum, mitochondria, among others, whereas bacterial cells are relatively simple regarding their internal structures. The different constructs between prokaryotic and eukaryotic cells relate to significant differences in biochemistry. For example, bacteria must synthesise many of the crucial molecules they require, such as essential vitamins, which animals obtain from food. Targeting these metabolic pathways unique to bacterial cells is one strategy for designing drugs to combat infections. Antimetabolite drugs selectively target bacteria; not affecting the host, by inhibiting enzyme-catalysed biochemical pathways that occur in the bacterial cell, but crucially not in the animal cells. Another mechanism by which antibiotics can be designed to combat bacteria is to directly target protein synthesis to prevent to the enzymes and other proteins that are essential to the bacteria from being made. Inhibition of nucleic acid transcription and replication is another way to disrupt protein synthesis and disrupt cell division.

Sulpha drugs are an example of antibiotics that affect bacterial cell metabolism. It was discovered in 1935 that a red dye called prontosil red was effective against bacteria *in vivo* (in other words in laboratory animals), but antibacterial effects were not observed *in vitro*; cultures in petri dishes were unaffected. It was later discovered that bacteria present in the small intestine of the animal metabolised prontosil to give the true antibiotic product called sulphanilamide. This process is illustrated in Figure 3.1. Prontosil was an early example of a pro-drug.

Sulphonamides proved effective against a wide range of infections and further developments extended the range to a number of gram-positive bacteria, especially pneumococci and meningococci. There were limitations to sulphonamides; they proved ineffective against *Salmonella*, and typhoid and problems arose in how the drugs were metabolised; often producing toxic by-products. Hence, sulpha drugs were eventually superseded by penicillin. Prior to the development of penicillin though, sulpha drugs were the antibiotics of choice for infectious diseases and are noted for saving Winston Churchill's life after falling ill during Second World War. Sulphonamides have been particularly useful against intestinal infections. Succinyl sulfathiazole (Figure 3.1) is a pro-drug of sulfathiazole; the succinyl moiety contains an acidic group, which becomes ionised in the weakly alkaline conditions in the intestine. As a result, it is not absorbed into the blood steam and is retained in the intestine. Slow enzymatic hydrolysis of the succinyl group releases the active sulfathiazole where it is needed.

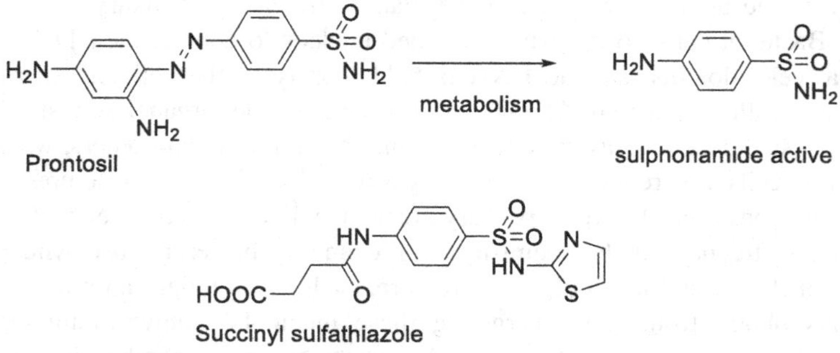

FIGURE 3.1 A Reaction Scheme Showing the Metabolism of Prontosil Red to Form the Sulphonamide Active (Above) and Succinyl Sulfathiazole (Below), An Example of a Sulphonamide Pro-Drug.

Structure-activity relationships were deduced from synthesis of a large number of sulphonamide analogues. The para-amino group must be unsubstituted because of their function; the only exception is where an acyl group is used, which is inactive, but is metabolised in the body to generate the active compound, in other words amides serve as a pro-drug. The aromatic ring and sulphonamide functional group are both required. The sulphonamide nitrogen must be primary or secondary and is the only site that can be varied.

A compound library of sulphonamide analogues are often varied by introducing heterocyclic or aromatic structures to the sulphonamide nitrogen, which effects the extent to which the drug binds to plasma proteins, and hence controls the blood levels and lifetime of the drug. A drug that binds strongly to a plasma protein will be more slowly released into the blood circulation and will be more long-lasting. Variations of the pharmacophore, SN(H)R (the Sulfur-Nitrogen(-Hydrogen)-Aromatic ring), where R can be different ring structures also alters solubility, so effects the pharmacokinetics of the drug, rather than its mechanism of action. For example, an N-acetyl group makes the compound insoluble, causing it to clog the kidneys and therefore is toxic. The compounds must contain an N-H proton, which has a low pK_a value, meaning that it is readily ionised and the drug becomes soluble.

The mechanism of action of sulphonamides is to work as competitive enzyme inhibitors and block the biosynthesis of folate in bacteria. Folate is an essential nutrient for bacteria, required for DNA replication. With restricted synthesis of folate, growth of bacterial populations subsides and the immune system can handle the infection. Humans acquire folic acid in food; bacteria are unable to do this and must synthesise their own, so there are no side effects to human health from deficiency of this nutrient. This kind of target specificity is referred to as the 'magic bullet' approach. By structurally mimicking the normal substrate for this enzyme, sulphonamides competitively inhibit it because the active site is tricked into accepting the drug, which therefore blocks access to the active site, albeit reversibly, to the substrate and reduces the production of folate.

Resistance can be acquired by bacteria that produce more of the normal folate than usual, thus competition for the active site favours the normal substrate. Alternatively, resistance can be acquired through mutations that alter the structure of the active site subtly to reduce the affinity for sulphonamides over the natural substrate. Mutations can also arise that decrease the cell membrane permeability to drugs and this is another issue regarding antibiotic resistance.

One very important difference between bacterial cells and those of animals is the presence of a cell wall that encapsulates the bacterial cell. This is in addition to a plasma membrane and serves to protect the cell, give it structure and stability, and, most crucially, to balance osmotic pressure;

otherwise, if bacteria without a cell membrane entered an aqueous environment containing a low concentration of salts, water would freely enter the cell due to osmotic pressure, which would cause the cell to swell and eventually burst. This is known as lysis. While the cell wall does not prevent water from entering the cell, it does stop the cell from swelling, and, therefore, indirectly prevents excessive amounts of water entering the cell, which would be catastrophic for the bacterium.

Bacteria can be classified according to their cell wall by using a staining technique. A purple dye is added, followed by washing with acetone. Bacteria with a thick cell wall (20–40 nm) absorb the dye and are stained purple. These are defined as gram-positive bacteria. Bacteria with a thin cell wall (2–7 nm) only absorbs a small amount of dye, which is washed out with acetone. These bacteria are then stained pink with a second dye and are defined as gram-negative bacteria. Although they have a thin cell wall, the important difference between these bacteria and gram-positive bacteria is that they have an outer cell membrane made up of lipopolysaccharides. This difference has important implications for the different vulnerabilities of gram-positive and gram-negative bacteria to antibiotics.

BREAKING DOWN BACTERIA'S DEFENCES

In 1877, Pasteur and Joubert discovered that certain moulds produced substances that are toxic to bacteria, but some of these were toxic to humans also, and therefore had no apparent application in medicine. In 1928, Fleming noted that a bacterial culture, which had been exposed to the air for several weeks, had become infected by a fungal colony. The area surrounding the fungal colony where the bacterial colonies were dying, he correctly attributed to the fungal colony producing an antibacterial agent. Realising the importance of this, he cultured the mould, which was found to be a relatively rare species of *Penicillium*; a spore of which must have originated in another laboratory and by good fortune blown on the wind into Fleming's laboratory. By chance the weather was just right for this to happen: a cold period allowed the fungus to grow while the bacterial population remained static; then a warm spell enabled the bacteria to grow and the antibiotic properties of penicillin to be observed. As a final twist, the culture plate was also stacked in a bowl of disinfectant above the surface of the sanitised water, ready to be washed up.

Through Fleming's keen observation, he was able to make this discovery and spent several years studying the novel antibacterial agent. He was able to demonstrate that this compound was non-toxic to mammals, but was not able to isolate and purify the unstable compound. This problem was solved by Florey and Chain in 1938, who used chromatography, and, by 1941, were ready to conduct the first clinical trials. The resounding success led to the commercialisation of penicillin and, by 1944, there was enough of it available to treat casualties from the D-day landings.

By this time, penicillin was in widespread use, but despite its popularity, chemists still debated its molecular structure, with many unusual characterisations being held in much scepticism. Surprisingly, it turned out that the structure of the compound was unusual. Dorothy Hodgkin established the structure by x-ray crystallographic analysis and revealed the β-lactam ring. The atoms in the ring are held tightly in a square structure, which is very strained and susceptible to breaking down, hence leading to Fleming's difficulty in isolating the compound.

Synthesis of such a highly strained compound was highly challenging, but manufacturing the compound without the need to laboriously purify it from mould cultures was advantageous for commercialisation of the antibiotic. This was accomplished by Sheehan in 1957. However, the full synthetic procedure was too involved to be commercially viable, but the following year a biosynthetic intermediate of penicillin, 6-aminopenicillonic acid (6-APA) was isolated by Beecham's and revolutionised the production of semisynthetic penicillin by producing a readily available starting material.

The four-membered β-lactam ring is unstable, being highly strained, and is fused to a five-membered thiazolidine ring in a bent shape. Its structure suggests that it is derived from the amino acids valine and cysteine. The nature of the acyl side chain R depends on the components of the fermentation medium.

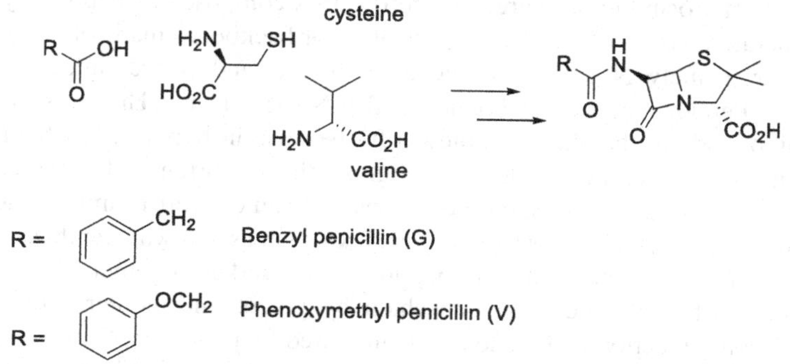

FIGURE 3.2 Overview of the Synthesis of Penicillin from the Amino Acids Valine and Cysteine.

The first mass production of penicillin used corn steep liquor as the fermentation medium, which contains high concentrations of phenyl acetic acid, and gives benzyl penicillin (penicillin G). Fermentation mediums containing phenoxyacetic acid give phenoxymethylpenicillin (penicillin V). These processes are summarized in Figure 3.2.

The properties of penicillin G make it a good choice to use as an antibiotic medicine. It is active against non-β-lactamase producing (in other words non-resistant) gram-positive bacteria, such as meningitis, gonorrhoea and several gram-negative cocci and anaerobic microorganisms, such as *Streptococcus* and *Enterococcus*, so is active against many rapidly dividing types of bacteria. Penicillin is non-toxic and very safe (magic bullet), however, some people have allergies to penicillin. Allergies range from a mild rash to anaphylactic shock. Small molecules such as penicillin generally do not cause this effect, but nucleophilic groups on proteins will react with the open β-lactam ring and therefore becomes covalently bonded to the protein, causing an immune response as the protein is 'recognised' as foreign. This only happens in 0.2% of patients. Other limitations to penicillin G is poor activity against gram-negative bacteria, so does not have a broad spectrum of activity. The molecule is broken down by stomach acid, so cannot be taken orally and must be injected, which is the least favourable route of administration to patients. To overcome some of these limitations, it was necessary to synthesis penicillin analogues. These compounds must retain the essential aspects of the molecule crucial to the mechanism of action, while possessing structural modifications that will change the interaction of the drug with the body in such a way as to overcome the limitations of the drug.

To explain the mechanism of action of penicillin, it is necessary to understand the structure of the bacterial cell wall. The cell wall is essential for bacteria to survive in a range of environments, such as varying pH, temperature, and osmotic pressure. The cell wall is composed of peptidoglycan; its structure consists of a parallel series of sugar backbones made of two types of sugar, namely N-acetylmuramic acid (NAM) and N-acetylglucosamine (NAG). Peptide chains are bound to the NAM sugars. The presence of D-amino acids in the chain is noteworthy because in humans, biochemistry only involves L-amino acids, which gives the opportunity for target selectivity. Bacteria have racemase enzymes that can convert L-amino acids to D-amino acids, required for biochemical uses such as cell wall synthesis. The cell wall is constructed with the peptide chains linked together by the displacement of D-alanine from one chain by glycine in another. This final cross-linking reaction is the one that is inhibited by penicillin. Consequently, the cell wall framework is no longer interlinked and is fragile; it does not stop the cell from swelling, which eventually bursts (lysis) and kills the bacterium. Different types of penicillin are used in this way to inhibit the transpeptidase enzyme responsible for cell wall construction for a range of bacteria.

The severity of damage is greater for gram-positive bacteria, which have thicker cell walls, consisting of 50–100 peptidoglycan layers.

The transpeptidase enzyme is classified as a serine protease; where a serine residue in the active site is involved in hydrolysis of peptide bonds. The serine acts as a nucleophile to split the two D-alanine units on a peptide chain. The terminal alanine departs, while the peptide remains in the active site. Another peptide chain enters the active site and a peptide bond is formed between D-alanine and the terminal glycine of the other chain. It is presumed that the penicillin conformation mimics the transition state conformation of the D-Ala-D-Ala moiety during the cross-linking reaction and the transpeptidase enzyme mistakenly binds it to the active site. The serine residue acts as a nucleophile and opens the β-lactam ring, but because the molecule is cyclic, it is not split in two as the peptide would be. Consequently, nothing leaves the active site, which is blocked and access to the second peptide chain is prevented. As a result, cross-linking in the bacterial cell wall is inhibited, making it fragile and lysis occurs.

Bacterial strains vary in their susceptibility to penicillin, and there are several ways through which they can derive resistance. In order to inhibit the transpeptidase enzyme, located on the bacterial cell membrane, penicillin molecules must traverse the cell wall, which is a rigid, but porous structure, so small molecules can readily pass. The cell wall does not present a physical barrier, however, gram-negative bacteria have an outer lipopolysaccharide membrane, which is impervious to polar molecules like penicillin G and water, and, hence, the lower activity of penicillin against them. Channel proteins called porins facilitate passage of water and essential nutrients across the cell membrane. The structure of this porin, as well as the nature of the penicillin molecule, determines the effectiveness of the drug against the bacteria. Typically, small hydrophilic penicillin analogues will more easily pass the outer membrane and be present in high enough concentrations to inhibit cell wall biosynthesis.

Some bacteria develop mechanisms to subvert the actions of penicillin. Mutated forms of transpeptidases result in the presence of β-lactamase enzymes, which open the β-lactam ring to form an ester bond as before, but are able to hydrolyse the ester link and expel the broken-down penicillin. Indeed, some gram-positive bacteria, such as *Staphylococcus aureus*, can release β-lactamase extracellularly and breakdown penicillin before it even reaches the cell membrane. Currently, 95% of *S. Aureus* strains release β-lactamase capable of hydrolysing penicillin G. This presents a huge problem with respect to treating *S. Aureus* infections.

Most, if not all, gram-negative bacteria produce β-lactamase, which is trapped between the cell membrane and outer membrane, and so is present at a higher concentration than in gram-positive bacteria; and hence the observed lower activity. However, variations in substrate specificity means that susceptibility to some penicillin types can occur in gram-negative bacteria. Efflux pumps in some gram-negative bacteria are able to remove penicillin from the periplasmic space between the cell and outer membrane. Resistance through mutation can be acquired quickly by bacteria as a consequence of their method of genetic transfer: fast cell division, called binary fission, and conjugation, where mutated genes are passed directly, speedily pass genetic information between populations.

With the surge of resistant strains of bacteria, it is essential that medicinal chemists produce a variety of penicillin analogues. Sheehan's full synthetic approach to making penicillin took too long and was too low yielding, and therefore was not commercially sustainable. This restricted medicinal chemist to fermentation and semisynthetic procedures. A limited number of variations were possible by adding different carboxylic acids to the fermentation process, thus changing the acyl group, but this method is time-consuming. Semisynthetic procedures involved fermentation to yield 6-APA, first isolated by Beecham's in 1959, when it was realised that this biosynthetic intermediate was the same as an intermediate in Sheehan's synthesis, which could be treated with a range of acid chlorides. Chemical methods involving hydrolysis of the acyl side chain is a more efficient method.

To consider the feasibility of a penicillin analogue, it is important to understand the structure-activity relationships of the drug. A strained β-lactam ring is essential. The free carboxylic acid is essential, as the penicillin is usually administered as a salt and the ionised carboxyl group binds to the charged ammonium ion of lysine in the active site. The bicyclic system is important to confer further strain on the β-lactam ring and thereby improve activity. Acylamino side chain is required; sulphur is usual, but not essential; and stereochemistry is important. In summary, very little variation is tolerated; modifications are limited to the acylamino side chain. Penicillin analogues needed to be produced that addressed the problems of acid sensitivity, β-lactamase susceptibility, and limited breadth of activity.

There are three reasons for acid sensitivity of penicillin: the strained β-lactam ring readily undergoes acid-catalysed ring opening, the β-lactam carbonyl is highly reactive with nucleophiles; not being resonance stabilised as with usual tertiary amines because mesomeric effects are impossible due to the increase in bond strain that would arise from a double bond being present within the already strained β-lactam ring (120 degree ideal vs 90 degree bond angle in the ring). Lastly, neighbouring group participation from the acyl side chain to open up the β-lactam ring: the self-destruct mechanism.

The mechanism in Figure 3.3 illustrates how the acyl side chain can be involved in the intramolecular ring-opening reaction. The deprotonated nitrogen atom acts as a nucleophile, attacking the carbonyl carbon with its lone

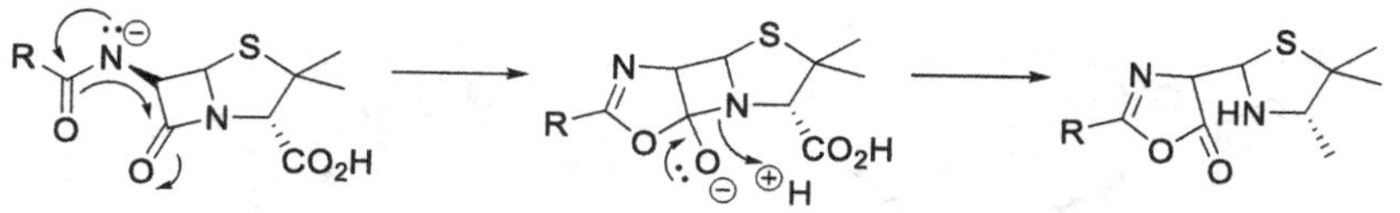

FIGURE 3.3 Mechanism for the Intramolecular Ring Opening of the β-Lactam Ring, Where Curly Arrows Show the Movement of Electrons Around the Molecule.

pair of electrons. Electrons from the carbon-oxygen double bond then move to the carbonyl of the β-lactam. Resonance of the electron pair in the β-lactam results in the carbon-nitrogen bond of the ring breaking and nitrogen forms a new bond to the proton that was lost to the solution initially.

To maintain an effective drug, the only modifications that can be made are to the acyl side chain, which can improve acid tolerance. To reduce neighbouring group participation, an electron withdrawing group is placed in the side chain to draw electrons away from the carbonyl and make it less nucleophilic, example the oxygen of penicillin V, therefore, this drug can be taken orally because it can survive the stomach acid.

Besides considerations for efficacy regarding the route of administration, modifications to the molecular structure of penicillin become important to combat antibiotic resistance. The need for innovative drug design to produce new antibiotics arises when the bacteria adapt and develop mechanisms that stop the drug from working. Penicillinase/β-lactamase resistant penicillin needed to be developed when, in the 1960s, widespread use of penicillin G led to an alarming increase in penicillin-resistant *S. aureus* infections.

The strategy of steric shields was implemented; in this approach, a sterically bulky group is placed on the acyl side chain, preventing access of the penicillin to the β-lactamase active site. However, design of a suitably sized steric shield was a challenge because medicinal chemists wanted to avoid compromising complementary fit between the penicillin molecule and the transpeptidase enzyme's active site. Thankfully, methicillin was developed just in time to address the growing number of cases of resistant *S. aureus* infections. With reference to Figure 3.4, the steric shields are the two ortho-methoxy groups on the aromatic ring that are able to make the discrimination between β-lactamase and transpeptidase active sites.

However, there are several drawbacks to methicillin: the absence of an electron-withdrawing group on the acyl side chain means that it is acid sensitive, it has only one-fiftieth the activity of penicillin G against micro-organisms that are sensitive to penicillin G, and shows poor activity against some streptococcal strains and is inactive against gram-negative bacteria. As a result, methicillin is no longer used clinically. Better penicillinase-resistant

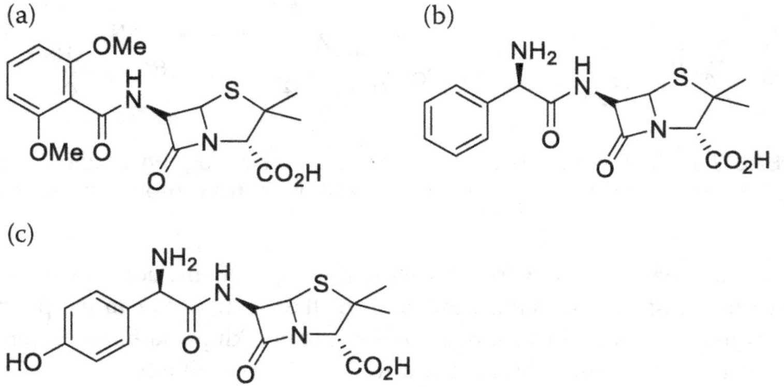

(a)   (b)

(c)

FIGURE 3.4   Molecular Structures for the Penicillin Analogues, (a) Methicillin, (b) Ampicillin, and (c) Amoxicillin.

penicillins have since been developed, but, disconcertingly, a large number of *S. aureus* strains that can be detected in hospitals have become resistant to methicillin and other penicillinase-resistant antibiotics as a result of mutation of the transpeptidase enzyme. These are termed methicillin-resistant *Staphylococcus aureus*, or MRSA.

With the perpetual occurrence of antibiotic resistance, the continued need for innovation requires medicinal chemists to work hard to develop new drugs to tackle the upsurge of resist pathogenic microorganisms. Different factors influence the efficacy of a drug and modification to the molecular structure of penicillin may or may not be successful in producing an effective drug. Factors effecting the spectrum of activity of penicillins include: the ability to pass the cell membrane of gram-negative bacteria, susceptibility to β-lactamase enzyme, affinity for transpeptidase target, and the rate of removal by efflux pumps. When searching for a novel compound, medicinal chemists must synthesise a library of compounds to find an analogue with the correct structure to overcome these issues.

It is found that hydrophobic groups on the side chain, as in the case of penicillin G, favour activity against gram-positive bacteria, but show poor activity against gram-negative bacteria. Analogues with hydrophilic groups (example NH_2, OH, and CO_2H) attached adjacent to C=O (α position) have little effect on gram-positive bacteria, but do enhance activity on gram-negative bacteria, example ampicillin and amoxicillin (Figure 3.4).

The modification of the penicillin by inserting an α-hydrophilic group enhances activity by aiding passage through the porins of gram-negative bacteria's outer cell membrane. These two analogues have a similar structure, but the additional phenol group improves absorption of amoxicillin through the gut wall. The NH_2 group produces acid tolerance, hence, these drugs can be taken orally. However, there are limitations, including sensitivity to β-lactamase (no steric shield), and they can cause diarrhoea, as poor absorption leads to high concentrations in the gut that can abolish healthy bacteria flora, enabling colonisation of harmful microbes, leading to intestinal distress.

The issue of poor absorption is a result of the dipolar nature of the molecule, which arises from the free amino acid group and carboxylic acid functionalities. The solution is to use a pro-drug approach, where one of the polar groups is masked by a protecting group that can be removed metabolically once the pro-drug has been absorbed. For example, the use of acyloxymethyl esters, which are susceptible to esterases. A nucleophilic residue in the active site of the esterase cleaves the carbonyl part of the ester from the pro-drug, and, in a second step, the penicillin molecule is unveiled. A curly arrow mechanism for this process is given in Figure 1 of the Supporting Material*. This can be any analogue mentioned previously, for example ampicillin protected in this way is called pivampicillin.

For these penicillin analogues, susceptibility to β-lactamase is an issue. However, when used in combination with clavulanic acid, the scope of antibiotics such as amoxicillin is greatly improved. Administered to the patient as Augmentin, the dose level of amoxicillin can be greatly decreased when used in combination with clavulanic acid because clavulanic acid is an irreversible inhibitor of most β-lactamases. It was first isolated from *Streptomyces clavuligerus* by Beechams in 1976.

Clavulanic acid was the first example of a naturally occurring structure with a β-lactam ring that is not fused to a sulphur-containing ring. Instead, it is fused to an oxazolidine ring. Also note the lack of acylamino side chain. Analogues have since been made, with the essential features for a β-lactamase inhibitor: the strained β-lactam ring, enol ether with Z-configuration, no substitution at C6, (R)-stereochemistry at positions 2 and 5, and a carboxylic acid group.

Drugs containing β-lactam rings are not the only chemicals that inhibit cell wall biosynthesis. Vancomycin is often referred to as the 'antibiotic of last resort' because it interrupts cell wall biosynthesis via a different mode of action to penicillin, so is effective against MRSA. The propensity for bacteria to develop resistance is why vancomycin is restricted to use as a last resort; when doctors have exhausted all other avenues, then this is the drug that is used. The hope is that restricting the exposure of bacteria to this drug will limit the risk of adaptation of the microorganisms to acquire resistance and hence prolong the effectiveness of vancomycin. Should bacteria evolve resistance to vancomycin,

---

*   Available at www.routledge.com/9780367644031

this would be severely detrimental to the arsenal that doctors have available to treat patients, and would promulgate innovation in the field of antibiotics. Vancomycin is a bacterial glycopeptide produced by a microorganism called *Streptomyces orientalis*. It is derived biosynthetically from a linear heptapeptide chain, containing five aromatic residues, which undergo oxidative coupling with each other to produce three cyclic moieties within the structure (see Figure 2 in the Supporting Material*). Chlorination, hydroxylation, and the final addition of two sugar units completes the structure. The cyclisation transforms a highly flexible heptapeptide molecule into a rigid structure, where the peptide backbone is held in one conformation and hindered bond rotation around the aromatic rings offers further stability.

The fixed conformation of the heptapeptide chain is necessary for the unique mode of action of vancomycin. The rigidity is necessary for the formation of five hydrogen bonds to the tail of a crucial pantapeptide cell wall building block. Dimerization occurs as a second vancomycin molecule forms four hydrogen bonds with the first. The steric bulk of these molecules prevents access to the pentapeptide cell wall component by transglycosidase and transpeptidase enzymes needed for cell wall biosynthesis. Being such a large molecule means that vancomycin cannot pass through the cell membrane, therefore is not effective against gram-negative bacteria.

Other antibacterial agents are designed to act on the plasma membrane. The polypeptides valinomycin and gramicidin A cause the uncontrolled movement of ions across the cell membrane. Valinomycin acts as an inverted detergent and complexes a 'naked' potassium ion. The hydrophobic outer of the complex can pass the cell membrane and deposit the potassium ion outside the cell, resulting in fatality. Typically, K^+ is in high concentration inside the cell and can only pass the membrane via specialised transport proteins and this equilibrium is disrupted by valinomycin. Valinomycin is selective towards K^+, having the correct spatial arrangement of donor atoms to displace water ligands and form bonds to K^+. Other ions, such as Na^+, are not the correct size to fit and displacing water ligands is too difficult. Unfortunately, the toxicity is not selective to bacteria; affecting mammal cells as well. Gramicidin A is a peptide consisting of 15 amino acids, which coil into a helix where hydrophobic side chains point outwards and interact with the membrane. Two helices of gramicidin A must combine to span the membrane and the hydrophobic interior of the helix serves as a channel for the passage of ions. However, gramicidin A is also toxic to humans. Producing compounds to serve as drugs which employ the 'magic bullet' approach is a challenge for medicinal chemists. Compounds must be toxic to bacteria, but safe to use and

---

*   Available at www.routledge.com/9780367644031

to achieve this, target specificity is essential when developing effective antibiotics in the future.

THE CONTINUED BATTLE AGAINST INFECTION

In the battle against pathogenic microorganisms, medicinal chemists have implemented other strategies of drug design besides producing compounds to target bacteria's defences. For example, antibacterial agents have been designed which impair protein synthesis.

Protein synthesis is orchestrated by the cell's DNA. An enzyme called DNA helicase separates the strands at a specific region on the DNA molecule; the gene containing the instruction for a specific protein, such as those like β-lactamase, which give rise to resistance. DNA helicase breaks the hydrogen bonds between the DNA bases, enabling another enzyme, called RNA polymerase to move along the template DNA strand and bind the exposed bases to complementary nucleotides that are present in the cell. This process is known as transcription and results in the production of a stand of RNA, which carries a complementary sequence of bases to the template gene on the DNA.

In eukaryotic cells, sections of the gene known as introns do not code for proteins and are removed during a process called splicing. The coding sections, called exons, can be arranged in different combinations, meaning that a single gene can code for several proteins. Once the final copy of the RNA, which is known as messenger-RNA, is made; this is produced directly from transcription in prokaryotes; it is available for translation, whereby the code contained in the sequence of bases is interpreted and the appropriate sequence of amino acids are constructed into a protein.

The sequence of bases on the mRNA are organises as discrete triplet codes; three bases code for one amino acid, and different sequences of triplet bases, each called a codon, code for a particular amino acid. For example, GAC codes for the amino acid aspartic acid. The process of translation is not as straightforward as the amino acids lining up along the mRNA strand. A second type of RNA, called transfer-RNA, is involved. The tRNA is a smaller molecule and is responsible for binding free amino acids in the cytoplasm and bringing them to the mRNA template. The tRNA molecules contain an anticodon, which is the opposite sequence to that on the mRNA and is complementary, therefore the tRNA carrying the amino acid can bind to mRNA.

Different codons are also present on the mRNA strand to determine where to start and terminate translation into a protein. Translation is done at organelles called ribosomes. A ribosome attaches to the starting codon on the mRNA molecule. A tRNA molecule, with a complementary anticodon, carries a specific amino acid and attaches to the mRNA by specific base pairing. A second tRNA molecule attaches to the next codon in the sequence in the same way. The ribosome moves along the mRNA molecule, working on two tRNA molecules at a time, and joins together the two amino acids via a peptide bond, using an enzyme and ATP. As the ribosome continues to move along the mRNA stand, the free tRNA molecules break loose and departs to collect another of the same amino acid from the pool of amino acids in the cytoplasm. This process continues until the ribosome reaches a stop codon (one that does not code for an amino acid) and the completed peptide is released. Note that up to 50 ribosomes can pass immediately after the first, so many identical polypeptides can be synthesised simultaneously. It is at this point that the tertiary structure would assemble from its constituent polypeptides.

Selective toxicity against bacteria can be achieved in drugs that target ribosomal RNA, and inhibiting different stages of the translation process, due to the fact that prokaryotic ribosomes differ in structure to those found in eukaryotic cells. The bacterial ribosome is a 70S particle, composed of a 30S subunit which binds to mRNA and initiates protein synthesis, and a 50S subunit, which binds to the 30S-mRNA complex to make the ribosome. The ribosome has two main binding sites: the peptide site (P site) binds the tRNA bearing the peptide chain, and the acceptor aminoacyl site (A site) binds the tRNA bearing the next amino acid in the protein sequence, to which the peptide chain will be transferred. Eukaryotic cells have bigger ribosomes made of a 60S large subunit and a 40S small subunit.

Streptomycin is an aminoglycoside: a carbohydrate structure which contains a basic amine group, and was the next most important antibiotic after penicillin. It proved to be the first effective agent against tuberculosis. Having being absorbed through the outer membrane of gram-negative bacteria, which was a limiting factor for efficacy of penicillin, streptomycin binds to the 30S subunit of the bacterial ribosome and prevents the triplet code on mRNA from being transcribed, so protein synthesis is terminated and vital proteins are not made.

There is a wide diversity of antibiotics designed to target bacterial protein synthesis. Tetracycline antibiotics are a broad spectrum class of antibiotics that also bind to the 30S subunit, but have a different mode of action. They prevent aminoacyl tRNA from binding and stop growth of the protein. Other antibiotics target the 50S subunit of ribosomes, such as chloramphenicol, and work by inhibiting the movement of ribosomes along the mRNA strand. Chloramphenicol is the drug of choice for treating typhoid in some parts of the world, where more expensive drugs cannot be afforded.

Other antibiotics target bacterial protein synthesis by acting on nucleic acid transcription and replication. Examples of the structures of this class of antibiotics are given in Figure 3 of the Supporting Material*. Quinolone compounds, such as nalidixic acid, are particularly useful for the short-term treatment of urinary tract infections. Resistance soon develops to these compounds; consequently new analogues had to be developed. Modifications to the structure of nalidixic acid were found to increase he spectrum of activity against both gram-negative and gram-positive bacteria.

A single fluorine atom at position 6 greatly increased activity as well as uptake into the bacterial cell. Addition of a piperazine ring on position 7 is beneficial; improved oral adsorption, tissue distribution, metabolic stability, as well as improving the level and spectrum of activity are among the advantages. Presumably, the ability for the basic substituent to form a zwitterion with the carboxyl group is the reason for these improved drug properties. Further modifications include addition of an isopropyl ring to nitrogen 1, and replacement of pyridine with benzene. This leads to the development of ciprofloxacin, which is regarded as one of the most active broad spectrum antibiotics available. Furthermore, bacteria are slow to develop resistance to it, unlike nalidixic acid.

The adaptive ability for bacteria to develop resistance is of great concern. A large proportion of *Streptococcus pneumoniae* and *Staphylococcus aureus* strains are resistant to β-lactam antibiotics. More ominously, resistance is even rising against the drug of last resort for treating MRSA, vancomycin. Hence, it is imperative that medicinal chemist continue to develop drugs to combat infection. Bacteria can acquire resistance against many antibiotics and this is known as multi-drug resistance. Multi-drug resistance can be acquired by the accumulation of multiple genes, each of which codes for resistance against an individual drug, typically on resistance plasmids within the cell. Multi-drug resistance can also occur from over expression of multidrug efflux pumps, enabling many types of antibiotic to be extruded from the bacterial cell. Some gram-negative bacterial strains are resistant to all known antibiotics, notably those belonging to *Pseudomonas aeruginosa* and *Acinetobacter baumanii*, and can serve as a reservoir for transmission of resistant genes. The emergence of so-called 'pan-resistant' bacteria coincides with a regression in novel drug development by major pharmaceutical companies, as the economic incentive to produce antibiotics is low compared to other drugs, such as anti-cancer agents. Society without these lines of defence to combat infection is a sombre thought.

There are several ways that resistance can develop. Bacterial cells divide and reproduce very rapidly; chance mutations that result in an enzyme or protein that impairs the effectiveness of a drug have a higher probability of

---

*   Available at www.routledge.com/9780367644031

occurring the longer the infection persists. Hence, it is imperative that a patient completes a course of antibiotic, even if symptoms of the infection subside long before the course is finished. The antibiotic needs to destroy the majority of the bacteria, then the body's immune system can cope with the remaining few that are more resistant. People failing to complete their course of antibiotics is a major factor contributing to upsurge in multidrug resistance.

Once drug resistance through chance mutation has arisen, other bacterial cells can acquire resistance through genetic transfer. Not only is the gene for resistance transferred to daughter cells as resistant bacteria divide, but also genes can be passed between bacterial cells. There are two main ways in which this can occur: conjugation and transduction. In conjugation, the genetic material is transferred directly between bacterial cells, through a connecting bridge of sex pili built between the two cells. Transduction involves small sections of genetic material, called plasmids, being transferred by means of bacterial viruses (bacteriophages) which may leave a resistant cell, then go on to infect a non-resistant cell with the relevant genetic material needed to acquire resistance; in by doing so, passing on the genetic information in the plasmid containing the instruction for resistance enzymes, such as β-lactamase.

Resistance is particularly prevalent in hospitals, where use of antibiotics is greatest. Trace amounts of drug are present in the air in hospitals and these have been attributed to development of resistance; as when breathed in they kill sensitive bacteria in the nostrils and thereby encourage proliferation of resistant strains. The prevalence of resistance is most widely attributed to the overuse of antibiotics. The carless use of antibiotics to treat minor infections in medicine, the widespread use of veterinary medicines including as additives to animal feed, all greatly enhance the possibility of resistance developing. Many bacterial strains are now resistant to the early antibiotics, such as penicillin G, yet in poorer third-world countries, where their use is more limited, these drugs are still effective.

The need to develop new drugs is essential to overcome the challenges of bacterial resistance. Sequencing genomes of proteins may elucidate new targets in the ever-present battle against pathogenic bacteria. Many of the compounds produced to combat pathogenic bacteria are semisynthetic, derived from microorganisms, and rely on fermentation for manufacturing them. By employing innovative research, targeting on a molecular-level approach might aid in designing novel compounds to battle infection, and overcome one of the leading challenges of medicinal chemistry today.

Antiviral Agents and Rational Drug Design

4

VIRUSES AND VIRAL DISEASES

Viruses must take over a host cell in order to survive and multiply. They are non-cellular infectious agents; essentially a protein packet containing nucleic acid, which is transmitted to the host cell to enable the virus to be replicated. Philosophically, it is debatable whether viruses can be considered as living entities, given that they do not produce their own energy and require living cells for their replication; although a predisposition for self-replication is a property of living things. Whether they may be considered as being corporeal, or just lifeless molecular entities, there are a large variety of viruses capable of infecting bacteria, plant, and animal cells; indeed over 400 different viruses are known to infect humans.

There are several ways that viruses can be transmitted. For example, airborne viruses, such as influenza, chicken pox, measles mumps, rubella, viral pneumonia, and small pox can be transmitted in sputum, when an infected person coughs or sneezes. Other viruses require physical contact to be transmitted because they cannot survive long outside a host, examples are the Herpes virus, which cause cold sores, HIV, and rabies. Viruses can even be transmitted by arthropods, such as biting ticks in the case of yellow fever and limes disease. Lastly, foodborne and water-borne viruses may be ingested, such as hepatitis A and E and viral gastroenteritis.

The impact that viruses have on society can be most dramatically demonstrated by historical events. It was speculated that smallpox was responsible for major epidemics that weakened the Roman Empire during the periods AD 165–180 and 251–266. The disease, brought home from travelling soldiers, would have raged for many years, indicated by historical records, killing many people whose immune system was not equipped to handle the disease. Smallpox was also responsible for the decimation of indigenous communities in North and South America, following first contact with Europeans during the early colonisation of the New World. The influenza pandemic in 1918 killed more people in this period than there were casualties in the First World War. It became known as Spanish flu because of censored press releases, to avoid loss of moral during the war; headlines reported cases in neutral Spain mostly, but illness and mortality in other European countries was rife.

In developing countries, viruses still present a devastating and omnipotent threat. Originally, these outbreaks were contained in small communities, and, although fatal to many in a localised remote area, the risk of widespread epidemic was smaller. However, with larger populations and the convenience of air travel, rare diseases are much more easily spread around the world. Scientist fear a catastrophic scenario involving the possible evolution of a 'super virus' with a transmission mode and infection rate akin to influenza, but with a much higher mortality rate. There are viruses that spread rapidly with high incidences of fatality, but fortunately the latency period between infection and detection of symptoms is short, so the diseases can be contained. However, should a virus evolve that has a latency period that is longer, more people would likely become infected and this would cause widespread devastation.

With contemplation of historical events, the containment and prevention of viral diseases spreading among populations is of crucial importance to ensuring safety in modern society. Instrumental in managing viral diseases are the medicines developed to combat the infections. Targeting viruses is very difficult, due to the considerable differences between their structure and that of cellular entities. In combatting bacterial infection, anti-bacterial agents could be found in nature and utilized through fermentation processes to produce antibiotics. Further modifications could be made to the structure to yield novel and effective semisynthetic drugs to combat resistance. Antiviral agents must be designed from scratch based on information from identified targets and this led to the development of rational drug design; a great advancement in medicine. Medicinal chemists now could not only incorporate natural products into designing new drugs, but design and synthesise completely unique chemical structures which interact with biological targets and offer a new line of defence against pathogens.

In order to be able to identify targets against which drugs can be developed, knowledge of the structure and life cycle of viruses is paramount. A virus particle can simply be considered as a protein package that contains a type of nucleic acid with which it can infect host cells and hijack the host cell's machinery to reproduce itself. Viruses contain one or more molecules of either RNA or DNA, but not both, thus are defined as either RNA viruses or DNA viruses. The RNA can be single-stranded (ssRNA) as is the case for most viruses, or double stranded, where the base sequence of the RNA that is the same as viral mRNA is called the (+) strand and its complementary partner is the (−) strand. Most DNA viruses contain the typical double stranded DNA, but single-stranded DNA is present in some viruses. There is great variation between viruses in the size of the nucleic acid, ranging from genomes coding for just three-to-four proteins to larger genomes that code for over one hundred proteins.

The protein packet, called the capsid, protects the nucleic acid. The capsid is composed of protein sub units, called protomers, which are manufactured by the host cell and aggregate spontaneously through self-assembly processes. An additional membranous layer may be present around the capsid, containing carbohydrates and lipids. The complete structure is known as the virion and these can range in size from 10–400 nm, so can only be viewed by electron microscopy.

Understanding the structure and life cycle of viruses is essential in devising strategies to combat viral infections by identifying suitable targets against which an antiviral drug can be designed. Since the structure of a virion is predicated around its regeneration, finding aspects of the virus' life cycle that can be impeded through intelligent designing of a drug would be hugely advantageous in the development of antiviral agents. Viruses require the host cell's capabilities of protein synthesis to manufacture the constituent parts of the virion, and this is how viruses reproduce themselves. To reiterate, by understanding the mechanisms of this process rational drug design can lead to the synthesis of an agent that will combat the virus.

Viruses display a typical lifecycle in which they infect a host cell, hijack the cell machinery to replicate themselves, then leave to infect other cells. The first stage is adsorption; a virion must first bind to the surface of the host cell by interacting with the surface glycoprotein receptors. In by doing so, the virus has tricked the cell into accepting the virus as a harmless entity. This leads to the second stage: penetration and uncoating, where the virus introduces its nucleic acid to the cell. This can be done by injecting it through the cell membrane, or by the virus entering the cell, then uncoating. Uncoating occurs in different ways: some viruses fuse with the outer of the cell membrane, then introduce the capsid containing the nucleic acid; other viruses enter the cell via endocytosis, where it is enveloped by the cell membrane, which pinches off to produce a vesicle containing the virion, which is brought into the cell. Enzymes within lysosomes then aid with uncoating.

Once inside the cell, the next stage is replication and transcription of the viral nucleic acid. The virus' genome is incorporated into the host cell genome, then the normal mechanisms for protein synthesis are commandeered; enzymes, ribosomes, amino acids etc. needed to manufacture proteins are utilized by the virus to produce the constituent parts of a virion, and because many proteins can be synthesised simultaneously, enough components are made to produce an inordinate amount of virus particles. These components arrange themselves within the cell to make new capsids through self-assembly. The final stage is virion release. Naked virions, with no outer membranes around the capsid, are released by cell lysis; bursting forth from the cell; leaving destructing in their wake. When the viruses acquire the final components to become mature, they move on to infect other cells.

ANTIGENS, IMMUNITY, AND VACCINATION

The body has in place defence systems to combat foreign invaders. The immune system comprises a class of white blood cells known as lymphocytes, which are present in large quantities in the blood and lymph fluid. These cells become activated when they detect foreign invaders, such as bacteria or viruses. Invading microorganisms are detected by lymphocytes from the unfamiliar molecules on the invader's cell surface, known as antigens.

The immune system has the ability to discriminate foreign molecules from 'self' molecules. Nearly any macromolecule that is foreign to the recipient has the potential to elicit an immune response and can be referred to as an antigen (antibody generator). The specificity displayed by the immune system; its ability to differentiate antigens is extraordinary; capable of distinguishing between two proteins that differ by only one amino acid, or even between two stereoisomers of the same molecule.

Immune responses can be broadly classified as either antibody responses, or cell-mediated immune responses. Antibody responses involve the production of antibodies by certain cells known as B cells, which circulate the blood steam and permeate other body fluids in search of the antigens that induced their creation. Antibodies are proteins known as immunoglobulins and they bind to this specific antigen, which inactivates its toxic effects by subverting the toxin's binding to cell receptors. Antibody binding to antigens on the cell surface of a pathogen marker the invading microorganism, making it readily identifiable by phagocytes; cells of the immune system that engulf and destroy the invader. Specialised cells, called T cells, that target the foreign invader, produced during cell-mediated immune responses, react with the antigens on the surface of infected host cells. The T cells may kill the virus-infected host cell, thus eliminating it before the virus has replicated, or in other cases the T cells may produce chemical messages to activate macrophages to destroy the invading virus.

Once the immune system has been subjected to a particular pathogen, it not only adapts to recognise the antigens from that pathogen, but develops a memory also, meaning that should this pathogen infect the body again, the immune system will react quickly to combat the infection. The antibodies produced that have high binding specificity to the antigens from this pathogen continually circulate in the blood ever after. The ability of the immune system to remember exposure to particular pathogens is utilised during vaccination.

Vaccination is a preventative approach to protect people from viral diseases and is particularly successful against childhood diseases, such as polio, MMR, small pox, and yellow fever. The approach was first realised by Edward Jenner in the 18th century, when he observed that a milkmaid, who contracted the less virulent disease cowpox, subsequently had become immune to smallpox. He inoculated people with material from cowpox legions and discovered that they too had become immune to smallpox.

Vaccination works by priming the immune system by introducing foreign material that has molecular similarity to some component of the virus, but lacks its infectious nature, or toxic effects. The molecular fingerprint of the virus is recognised as 'non-self' and the cells of the immune system prepare their defence so that, should the virus infect the body, the immune system is ready to attack it. Generally, a weakened or dead virus is administered, or components of the virus that display a characteristic antigen (subunit vaccine).

However, there are limitations to vaccination. Vaccines usually are not effective on patients that have already become infected by the virus. Also, rapid gene mutation in viruses constantly changes the amino acid composition of glycoproteins on the virus surface, which leads to different antigens that are not recognised by the immune system, so the virus is disguised. Vaccines may not be suitable for patients with a weakened immune system, such as individuals with cancer, an organ transplant, or HIV, therefore another strategy is needed: antiviral drugs.

The reason that vaccination generally does not work on patients where infection has already occurred is because, for most of its life cycle, the virus is within a host cell and so is hidden from the immune system and circulating medicines. Moreover, because the virus capitalises on the cell's own biosynthetic mechanisms, the options for potential drug targets are limited compared to other invading pathogens. This makes developing antiviral drugs challenging.

Medicinal chemists are presented with the challenge of devising a strategy to combat viruses and rationalise and design drugs for any potential target. The design of such drugs will be cognisant, based on research into the virus structure and life cycle. Until the 1980s, very few clinically useful drugs were available for treating viral infections. Since then, progress has accelerated due to the AIDs pandemic and improved understanding of viral infectious mechanisms from viral genomic research, which has propagated great advancements in this area. Indeed, a full viral genome can now be quickly determined and compared with those of other viruses, enabling identification of how the virus' genetic sequence is split into genes. While genetic sequence will vary from one virus to another, it is possible to identify similar genes that code for proteins with similar functions, and thus present a possible target.

Genetic engineering methods can be implemented for the production of pure copies of the protein by inserting the viral gene into a bacterial cell, so that sufficient quantities of the protein can be synthesised and isolated to be used for screening. The viability of the protein as a potential drug target can then be determined by studying protein-drug interactions. Proteins that are good drug targets must fulfil certain criteria. They must be crucial for the life cycle of the virus, therefore have a major effect on the mode of infection. The proteins must bear little resemblance to human proteins, therefore the drug would be expected to have good selectivity and minimal side effects. Ideally, the protein will be common among a variety of viruses with a region of amino acid sequence that is identical and conserved, therefore the drug would have a broad spectrum of activity. Importance for an early stage of the virus life cycle is advantageous, then the virus is prevented from spreading throughout the body and causing symptoms.

ANTIVIRAL DRUGS IN ACTION

Finding a target to combat viruses is a challenge. A prudent strategy might be to investigate the viral genome and attempt to develop a drug that targets the virus's nucleic acid, given that this is the critical part of the virus, needed for its replication. Different viruses have different nucleic acids, so rationally designed drugs to combat one type of virus may not be effective against another.

Nucleic acids are polymers composed of nucleotide monomers; these building blocks of DNA and RNA are composed of a nitrogenous base (purines or pyrimidines) attached to a sugar molecule and a phosphate group, which make up the backbone of the nucleic acid. The sugar molecule bonded to a base can be referred to separately as a nucleoside.

Most drug development against DNA viruses are targeted against Herpesviruses to tackle diseases such as cold sores, genital herpes, chickenpox, shingles, as well as Burkett's lymphoma and Kaposi's sarcoma. Acyclovir was discovered by compound screening and brought onto the market in 1981. It revolutionised the treatment of these diseases, being the first relatively safe, non-toxic drug to be used systemically for the treatment of these diseases. Acyclovir is a nucleoside analogue; structurally similar to the nucleotides that make up DNA and contain the same nucleic acid base as deoxy guanosine, but lacks the complete sugar ring.

Acyclovir, shown in Figure 4.1, is in fact a pro-drug; the active agent is generated by phosphorylation in three stages to from a triphosphate within the infected cell. Nucleotide trisphosphates are the constituent building blocks of DNA, which assemble together along a template strand through the action of enzymes during DNA replication. Namely, DNA polymerase is responsible for this process and acyclovir works as an inhibitor of viral DNA polymerase and prevents DNA replication in two ways: either by inhibiting DNA polymerase, or as a chain terminator. DNA polymerase can catalyse the attachment of acyclovir into the growing DNA chain because it has a sufficiently similar structure to deoxy guanosine. Since the sugar ring is incomplete, and lacks the required hydroxyl group on position 3' of the sugar ring, the nucleic acid chain cannot extend further and DNA replication ceases.

DNA replication is essential in uninfected cells also; required to replenish healthy tissue, so the risk of side effects needs to be considered carefully. However, selectivity arises as acyclovir is only converted to the active triphosphate in infected cells. Although the enzyme that catalyses phosphorylation is present in normal cells, the Herpesvirus has its own version of the enzyme, which catalyses the phosphorylation of acyclovir more rapidly than host cell enzymes. Therefore, in healthy cells, acyclovir remains as the inactive pro-drug.

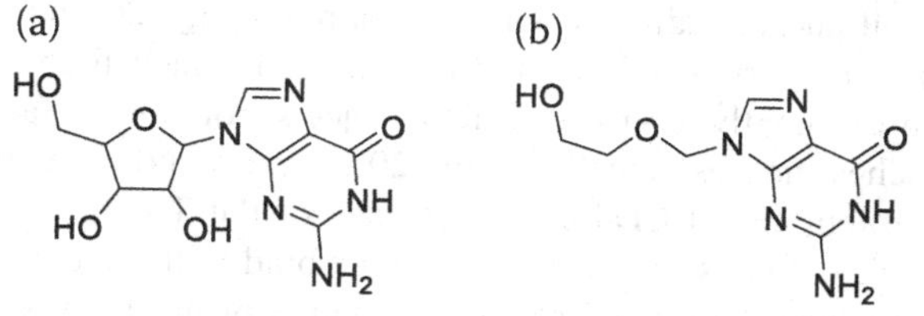

FIGURE 4.1 Structures of the Nucleic Acid Base Guanosine (a) and the Nucleoside Analogue Acyclovir (b).

A major limitation of acyclovir is the low bioavailability (15–30%) when taken orally. Pro-drug analogues were developed that remedied this issue by introducing groups to make the compound more water-soluble. Valacyclovir has an L-vinyl ester group that is hydrolysed in the liver or gut wall to unveil acyclovir, which now has a blood level concentration equivalent to intravenous injection.

Exasperatingly, strains of Herpes are appearing that show resistance to acyclovir. This is on account of mutations arising in either the enzyme responsible for phosphorylation of acyclovir, or in viral DNA polymerase. The consequences of these mutations are that the active drug is not generated in the cell, or the active drug is not recognised by viral DNA polymerase and replication of the virus continues uninterrupted.

A very relevant case study to examine drug development against RNA viruses is the treatment of human immunodeficiency virus, HIV. Acquiring this virus causes patients to have a weakened immune system and, without medicine, can lead to acquired immune deficiency syndrome (AIDS), which is usually fatal. Patients with AIDS become fatally susceptible to secondary infections such as pneumonia and fungal infections, which can escalate and lead to death. Scientists discovered that the immune response in patients with AIDS had been weakened by a virus, HIV. The virus infects T cells, which are crucial to the immune system and therefore directly attacks the immune response, meaning that patients are less able to cope with secondary diseases.

HIV is an example belonging to a group of viruses known as retroviruses; there are two variants associated with geographical locations: HIV 1 is responsible for AIDS in American, Europe, and Asia, HIV 2 occurs mainly in Western Africa. The extent of prevalence of AIDS has propagated a great deal of research into HIV and currently antiviral drugs have been developed to act against two viral enzyme targets: reverse transcriptase and protease. However, these drugs only mitigate the proliferation of the disease and do not eradicate it; hence, ongoing research is directed at finding better targets.

HIV is an RNA virus containing two identical (+)-ssRNA strands in the capsid, along with viral enzymes. The capsid is enclosed by a layer of matrix protein. This virus particle is enveloped by a membrane originating from that of the host cell and contains the viral glycoproteins gp120 and gp41, which are crucial for adsorption and penetration. The gp41 spans the membrane and is bound non-covalently to gp120, which projects from the surface. When the virus approaches the host T cell, the gp120 interacts and binds with a transmembrane protein called CD4 on the surface of the T cell. Conformational change of gp120 unveils gp41 which can now bind to the T cell, and anchors the virus to its host, then conformational change of gp41 pulls the virus towards the T cell so that their membranes can fuse and allow penetration of the HIV capsid.

Once inside the host cell, the protein capsid disintegrates with the assistance of viral protease enzymes. The contents of the capsid are then released into the cell cytoplasm. The viral RNA is not capable of coding for proteins and self-replication directly, but with the assistance of viral reverse transcriptase, is converted into viral DNA and incorporated into the host cell DNA. The virus can now remain dormant until cellular processes promote transcription of the viral genes within which the instructions for making the necessary matrix proteins, glycoproteins, and crucial viral enzymes are encoded. The viral glycoproteins gp120 and gp41 are produced and incorporated into the cell membrane of the host. As these proteins are manufactured, the constituents of the virion gather together at the membrane, then budding releases the components of the virus. The constituent proteins then self-assemble to produce the virion.

Viral genomics and other research has enabled scientists to gain an understanding of the lifecycle of HIV and thus identify suitable targets, namely reverse transcriptase and protease, and develop drugs that interfere with their function. However, the occurrence of mutation is very high in HIV and acquisition of resistance is a big problem. Long term, this can result in selection of mutated viruses that are resistant. As such, using a multidrug approach, where a combination of therapies are administered, tends to be the most efficacious strategy. This method has been successful in delaying the progression to full AIDS, but is not a cure.

There are three types of drugs available for highly active antiretroviral therapy (HAART), which may be used in combination: nucleoside reverse transcriptase inhibitors (NRTIs), non-nucleoside reverse transcriptase inhibitors (NNRTIs), and protease inhibitors (PIs). These drugs are largely synthetic in nature, but are based on natural product structures, such as modified arabinose nucleosides, in particular the protease inhibitors which except for one, are isosteres of the native hexamer peptide substrate. To be efficacious as a therapy, these drugs must fulfil certain requirements. They must have high affinity and selectivity for their targets and thus prevent the virus from multiplying and spreading, while being safe and well tolerated. They must be synergistic with each other so that they can be used in a combination therapy, but also compatible with other drugs that might need to be used to cure secondary infections, such as antibiotics for pneumonia. They should be suitable to take orally, with minimal dosing frequency to maintain therapeutic concentrations in the circulation, because they are likely to be required for the duration of the patient's lifetime. Preferably, the drug will be able to pass the blood-brain barrier in case the virus resides within the brain.

The enzyme reverse transcriptase is a DNA polymerase exclusively associated with the virus, but caution should nevertheless be taken to ensure that there is no inhibitory effect on cellular DNA polymerases when designing nucleoside reverse transcriptase inhibitors. These molecules mimic the nucleoside structure and become phosphorylated to the active drug, as explained in the previous example, acyclovir. The important difference here though is that the phosphorylation steps must all be carried out by cellular enzymes because, unlike Herpesvirus, HIV does not have the necessary kinase enzyme.

The first drug to be approved for use against AIDS was zidovudine; a deoxy thymidine mimic that was originally developed in the 1960s for treatment against cancer. The use of anti-nucleoside drugs as antiviral medications is rational, with clearly understandable mechanisms of action in preventing DNA replication. The sugar 3' hydroxyl group is replaced by an azido group; it inhibits reverse transcriptase, but also, on account of the azido group on the sugar ring, it acts as a DNA chain terminator. Originally purposed as an anticancer agent, zidovudine was designed to stop the proliferation of cancerous cells by disenabling DNA replication needed for cell division. As might be expected for such a compound, zidovudine causes severe side effects, which ultimately led to its use being discontinued.

Indeed, the severe side effects of zidovudine, also known as AZT (azidothymidine), meant that the drug was very controversial during the late 1980s/early 1990s when it was approved for use against HIV/AIDS. At the time, people did not appreciate that AZT was a breakthrough in the treatment of HIV/AIDS and ultimately helped to destroy the notion that the disease was a death sentence. Since it was first realised that HIV was the cause of AIDS in 1984, concerted efforts to rationally design antiviral agents led to the utilization of AZT, which was shown to be effective against HIV *in vitro* in 1985, and shortly after was approved for use against full AIDS in 1987, then approved for HIV in 1990. This was one of the fastest routes to clinic for any drug in the modern era. Campaigners argued that the early closure of the clinical trials was evidence that AZT was dangerously toxic and that this is why it failed cancer trials. This of course was not true; it is standard practice to give all patients a therapy once it is shown to save lives in a statistically significant way.

Besides the problems of associated side effects, viral resistance to AZT also became a problem, particularly if AZT was not used as part of a combination therapy. Consequently, it became a matter of urgency to develop new anti-HIV drugs, designed to avoid the problems of AZT resistance and improve on its performance. This led to a range of nucleoside-based reverse transcriptase inhibitors being introduced to the market. Understanding of the structure-activity relationships with the target enzyme enable less-toxic analogues to be developed, such as lamivudine, where the 3'-carbon is replaced with a sulphur atom which serves as a chain-terminating group. The structures of AZT and lamivudine are compared with the nucleoside thymidine in Figure 4 of the Supporting Material*.

---

* Available at www.routledge.com/9780367644031

Non-nucleoside reverse transcriptase inhibitors have a different mode of action. The molecules are generally hydrophobic and bind to a hydrophobic allosteric binding site; a binding site on the enzyme that is different to that used by the normal substrate, namely the active site. The physical effect of binding to a protein changes its shape, therefore deforms the active site, rendering it unusable by the normal substrate. NNRTIs are referred to as non-competitive reversible inhibitors because they are not in conflict with the normal substrate for access to the active site and their binding to the reverse transcriptase is not permanent. These two binding sites are distinct, so NNRTIs can be used in combination with NRTIs, which is advantageous because mutation can rapidly change the allosteric binding site and without a multidrug approach would result in the preservation of resistant viruses.

The first-generation NNRTIs were discovered through random screening programmes. Neviroprine has a rigid butterfly-like conformation making it chiral, where one 'wing' forms hydrophobic and van der Waals interactions with aromatic residues in the binding site, while the other wing interacts with aliphatic residues. Another first-generation drug is delavirdine, which was developed from a lead compound that emerged from a screening programme of 1500 structurally diverse compounds. Its structure differs from that of neviroprine most noticeably by the large tail that extends outside the normal binding site and projects out into the surrounding solvent. The alkyl pyridine ring motif is preserved, where hydrophobic contacts are important for binding to the same amino acids in the binding site. Unlike other first-generation NNRTIs, there is hydrogen bonding to residues in the main peptide chain of the binding site.

Changes to the amino acids in the binding site that form non-covalent interactions with the drug led to resistance against the first-generation NNRTIs. Crystallographic studies of the binding interactions and amino acid sequencing revealed that often a large amino acid is replaced by a smaller one, associated with the loss of an important binding interaction. Second-generation drugs were rationally designed to combat resistant viruses. Efavirenz is one such second-generation drug, developed as a result of comprehensive research on structure-binding relationships deduced from x-ray crystallographic interrogation of drug-binding site complexes of the first-generation drugs. X-ray crystallography and molecular modelling led to the structure-based design of a series of protease inhibiters. Multi-drug approaches, using a combination of reverse transcriptase inhibitors together with PIs is the most successful method of treating HIV infections because activity is improved and preservation of resistant mutations is less likely.

Reverse transcriptase inhibitors are pro-drugs and are activated by phosphorylation in vivo, but this is not the case with protease inhibiters. This is advantageous because it means that *in vitro* assays can be performed, where antiviral activity can be tested on infected cells in a petri dish, without the need for a patient or subject to be recipient of the untested drug, which has obvious ethical implications. Also, the protease enzyme can be isolated and the PIs can be studied directly by x-ray crystallography to determine structure-activity relationships, therefore an effective drug can be readily designed.

In this way, important tests can be carried out without a risk to life and an effective drug can be designed, then used to carry out the necessary tests, such as IC_{50}, which gives an indication of how effective the protease inhibiter will be. The IC_{50} is the blood concentration of the drug required to inhibit 50% of the enzyme, so the lower this value is, the more potent the drug. However, a strong inhibiter does not necessarily correspond to good activity. In order to be effective, the drug must be able to pass the cell membrane of infected cells. Hence, this is why *in vitro* whole cell assays are used alongside enzyme studies to ensure good absorption. Isolated lymphocytes infected with HIV are treated with the novel drug, antiviral activity is measured and recorded as an EC_{50} value; the concentration required to inhibit 50% of the cytopathic effect of the virus.

Good oral bioavailability is another important issue because anti-HIV drugs need to be taken for the duration of the patients' life, so a convenient route of administration is imperative. This is a particularly problematic aspect of PI development. PIs are developed from peptide lead compounds. Peptides notoriously have poor pharmacokinetic properties; poor absorption, metabolic susceptibility, rapid excretion, limited access to CNS, and high plasma protein binding. This is due to high molecular weight, poor water solubility, and the susceptibility of peptide bonds to hydrolysis.

The HIV protease enzyme is an aspartyl protease, which contains an aspartic acid residue in the active site, which is crucial for the catalytic cleavage of peptide bonds. The enzyme is a relatively small protein that can be readily made by synthetic techniques or by cloning and expression in rapidly dividing cells, then isolated and purified in large quantities. Crystallisation of HIV protease is relatively straightforward, hence this enzyme is an ideal target for rational structure-based drug design. From x-ray crystallographic studies, novel inhibiters can be developed to produce promising lead compounds.

The HIV protease enzyme is a dimer composed of two identical protein units, each consisting of 99 amino acids. The active site resides at the interface of the two units and, like the overall protein structure, is symmetrical about an axis of two-fold rotational (C2) symmetry. It has a broad substrate specificity; being able to cleave a variety of peptide bonds in viral polypeptides, but crucially it will hydrolyse bonds between proline and an aromatic residue (phenylalanine or tyrosine). This is an important feature because cleavage of these bonds is unusual and cannot be done by mammalian proteases. Furthermore, mammalian proteases lack the C2 symmetry characteristic of HIV protease. These features offer the possibility of drug selectivity.

HIV protease can be readily synthesised or isolated from cellular assays, then easily crystallised for study under x-ray diffraction and molecular modelling, as well as the differentiation from mammalian proteases, makes this a good target for anti-HIV drug design. The design of HIV protease inhibiters was inspired by the extensive research done on the mammalian aspartyl protease, renin. This enzyme was studied prior to the discovery of HIV protease and has a similar hydrolytic mechanism, hence its inhibiters have a close resemblance to those of HIV protease.

These agents act as transition state inhibiters, which mimic the transition state of the substrate during this enzyme catalysed reaction. This approach is advantageous because the transition state structure is more strongly bound to the active site than the substrate or products, so inhibiters that are structurally similar to the transition state will be most effective. As shown in the reaction mechanism in Figure 4.2, the transition state resembles the intermediate (2). However, the structure is inherently unstable, so an inhibiter must be designed that contains a transition state isostere, which mimics the tetrahedral centre of the transition state, but is stable to hydrolysis.

The importance of the dimer nature of the protease enzyme is evident in these mechanisms, as the two equivalent aspartate residues, Asp-25 and Asp-25' work cooperatively to catalyse the hydrolysis of the amide bond of the isostere shown explicitly here without convolution from drawing the rest of the substrate structure. This is an acid-base mechanism, using water as a nucleophile, which is activated by the Asp-25' residue. The TS intermediate (2) is an unstable structure, so the challenge for medicinal chemists was to incorporate this tetrahedral TS isostere into a molecular scaffold to produce a stable compound that would perform as an effective drug.

Several of these isosteres had been developed previously in the design of renin inhibiters. As a result, a huge library of compounds were synthesised incorporating these isosteres, with hydroxyethylamine proving to be the most successful. The effectiveness of this isiotere can be attributed to the hydroxyl-substituted amine bearing close resemblance to the TS intermediate, and interacts with the aspartate residues in the active site in much the same way.

The next stage was to design protease inhibiters based on the enzymes natural peptide substrate and to incorporate the isostere at the correct position. The first PI to be developed was saquinavir, by Roche. The natural peptide substrate was studied and the regions containing a phenylalanine-proline link were located, then a peptide sequence Leu[165]-Asn-Phe-Ile[169] was identified and served as the basis for inhibitor design. Since it is actually the phenylalanine-proline peptide bond that is hydrolysed, this link was replaced by the hydroxyethylamine TS isostere. Consequently, the usual hydrolysis reaction cannot occur and the substrate functions as a successful inhibiter. Additionally, the Leu-Asn-Phe-Ile residues are preserved in the substrate in order to bind

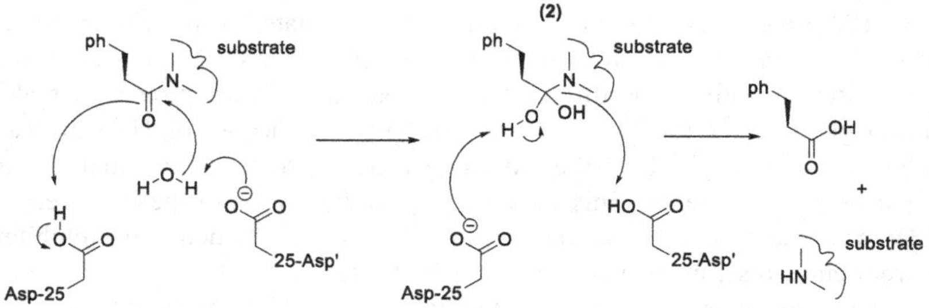

FIGURE 4.2   Mechanism of the Enzymatic Hydrolysis of the Transition State Isostere.

effectively to protease subsites. Despite this, inhibition of the enzyme is relatively weak. Furthermore, high molecular weight and peptide-like characteristics are detrimental to oral bioavailability. As a result, Roche needed to
find a smaller inhibiter based on the TS isostere that replaces the Phe-Pro
dipeptide. An N- and C-protected structure of the hydroxyethylamine was
tested first and was found to have weak inhibitory activity. The inclusion of an
asparagine group to occupy one of the subsites resulted in a 40-fold increase in
activity; greater than that of the original pentapeptide. This may seem para-
doxical, as the latter occupies four binding subsites, but it has been found that
the crucial interaction is that of the asparagine with its binding subsite. Other
binding interactions to other subsites weaken this core interaction, therefore
there is a drop in activity.

From these findings, it was concluded that the hydroxyethylamine isostere bonded to asparagine had the greatest efficacy and was selected as the
lead compound which led to saquinavir. X-ray crystal analysis revealed that
the protecting group occupies another subsite, which was shown to be a large
hydrophobic pocket. The protecting group was replaced by a large quinolone
ring system to optimise binding to this pocket and this resulted in a six-fold
increase in activity. Further modifications to the carboxylate end of the molecule yielded saquinavir (Figure 4.3), which had a further 60-fold activity
compared to the prior lead compound.

Contemplation of the efficacy of the drug can be perceived from the
properties of saquinavir. It is selective for HIV-1 and HIV-2 proteases; one
hundred times greater than for human proteases. Clinical trials were conducted
in 1991 and the drug reached the market in 1995, indicative of the faith that
medicinal chemists had in this medicine. However, about 45% of patients
developed clinical resistance over a one year period. Resistance is delayed if
taken in combination with reverse transcriptase inhibitors, though. Oral bioavailability is poor; only about 4% in animal studies, but this can be improved

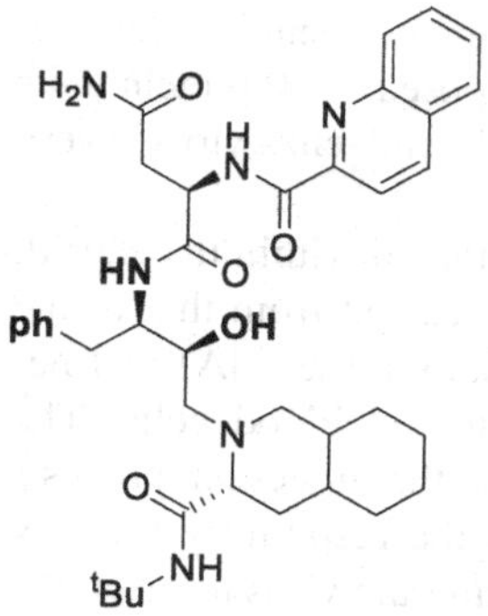

FIGURE 4.3 Molecular Structure of Saquinavir, an Anti-HIV Drug, with the TS Isostere Shown in Bold.

when taken with meals. There is also 98% binding with plasma proteins, therefore the drug needs high dosing to achieve therapeutic plasma levels.

While saquinavir was an advancement from previous lead compounds, a more efficacious drug was required, with a lower molecular weight, less peptide character, and better oral bioavailability. Merck had designed a potent PI by taking advantage of the symmetrical nature of the protease active site. By combining half of one PI with half of another, it would be possible to create a structurally distinct hybrid inhibiter. Chemists at Merck decided to combine the p' half of their lead compound, which has less peptide character, with the p' half of saquinavir because of its solubility enhancing potential. Lead optimisation led to the structure of indinavir, which reached the market in 1996. It had improved oral bioavailability (15%) and is less highly bound to plasma proteins (60%). This approach of combinatorial chemistry was extended to produce other ranges of PI inhibiters also.

Influenza is another example of an RNA virus; the rational design of drugs to treat this pathogen yielded different strategies to combat viruses to those used against HIV. Influenza causes the airborne disease commonly known as the flu when it infects the epithelial cells of the upper respiratory tract. It is a major source of mortality, particularly among the elderly and persons with a weakened immune system. Most notable is the pandemic in 1918, where Spanish flu caused the death of at least 20 million people worldwide; greater than the number of casualties in the First World War. Further serious epidemics have since occurred, and it is likely that their origin is from China, where people live in close proximity to poultry and pigs, so the risk of cross-species infection is much higher.

Occasionally, and mistakenly, the terms 'flu' and 'cold' are used interchangeably, example the severe disease called 'man flu'. Note that the common cold is caused by a different kind of virus to influenza, called rhinoviruses, so different therapeutic agents are required to combat colds. Human rhinoviruses are among the smallest animal RNA viruses, containing a single strand of positive RNA enveloped by an icosahedral capsid composed of 60 protein units of which there are four distinct types. The structure of the influenza virus is very different.

The (−) single-stranded RNA of influenza is contained within the capsid, which itself is enveloped in an outer membrane constructed from that of the host cell and contains two viral glycoproteins: neuraminidase (NA) and hemagglutinin (HA); so called because it can bind to red blood cells. The function of these two glycoproteins is to facilitate the infection process. NA helps the virus to traverse the mucus layers of the respiratory tract by catalysing degradation of the mucus layer, thus enabling the virus to reach the surface epithelial cells. Once there, adsorption of the virus can occur, whereby the virus binds to the host cell receptors that are recognised by HA, which binds to them rather than catalysing their degradation. The virion is now adsorbed; this process is known as receptor-mediated endocytosis. The pH then decreases inside the endosome, which is the membrane-bound compartment inside the host cell that contains the virion, causing HA to drastically change its conformation, where the hydrophobic ends of the protein fold outwards; extending towards the endosomal membrane and fusion occurs on contact, enabling the RNA capsid to be released into the cytoplasm of the host cell. Disintegration of the capsid releases the RNA and the viral enzyme RNA polymerase; both of which invade the nucleus.

Viral RNA polymerase now begins to catalyse the copying of the viral (−) RNA as (+) viral RNA, which departs from the nucleus and acts as the mRNA needed for the translation of viral proteins. Capsid proteins made in the cell spontaneously self-assemble in the cytoplasm with incorporation of replicated strands of (−) RNA as well as newly synthesised RNA polymerase. The glycoproteins NA and HA are then incorporated before the virion is released from the cell by budding. Interactions of the glycoproteins NA and HA are crucial to this step. There needs to be a balance between the rate of desialylation by NA to aid the virion's departure from the cell and the rate of binding by HA, which was required for access to the cell. The amino acids present in the active site of NA are highly conserved, which illustrates the importance of the activity level of this enzyme.

The glycoproteins NA and HA are present on the outer surface of the virion, so can serve as antigens; potentially recognised by antibodies and the cells of the immune system. While the amino acids in the active site of NA and HA are highly conserved, the influenza virus is adept at varying the amino acid sequence elsewhere in the proteins: this propensity for mutation hinders recognition by the immune system and limits the ability to develop vaccines with these antigens. The reason for this high level of antigenic variation is because the RNA polymerase enzyme is highly error prone. This results in the production of RNA that codes for NA and HA being inconsistent. Variation in the RNA code means that different sequences of amino acids give different antigenic variants, resulting in different strains of flu. Where the variation is small, it is referred to as antigenic drift. When there is large variation, called antigenic shift, this can lead to more serious epidemics and pandemics.

As mentioned previously, developing a vaccine for influenza can present a challenge. For eventualities where vaccination proves unsuccessful, medicinal chemists must develop antiviral agents which remain effective despite antigenic drift. Given that neuraminidase is crucial to the infectious process, it presents a promising target for drug development.

Studies of the enzymes crystal structure using x-ray diffraction and molecular modelling revealed that neuraminidase is a mushroom-shaped tetrameric glycoprotein, attached to the viral membrane by a single hydrophobic sequence consisting of 29 amino acids, which can be cleaved from the surface enzymatically to enable the unadulterated polypeptide to be studied without loss of antigenic or enzymatic activity. It was found that the active site is a deep pocket located centrally on each protein subunit; composed of 18 amino acids that are highly conserved. The outer structure of the protein is much more variable.

Since the amino acids in the active site are constant, and inhibition of the active site would severely affect the infectious process, neuraminidase inhibiters are an attractive target for medicinal chemists to create an antiviral therapy against influenza. By finding an inhibiter for the active site of NA, all strains of influenza would be targeted. Furthermore, the neuraminidase active site is unlike those of comparable enzymes for bacterial or mammalian cells, so there is no complication with selectivity.

During efforts directed towards elucidation of a structure for a possible inhibiter; NA has been crystallised with the product of the enzyme-catalysed reaction, sialic acid, bound to the active site in order to study the interactions. The most important interactions are hydrogen bonds and ionic interactions or the carboxylate group of sialic acid with three arginine residues, particularly with Arg-371. To achieve these interactions, sialic acid needs to be distorted from the most stable chair conformation, where the carboxyl group placed in an axial position on the chair changes to a pseudo-boat conformation, where the carboxylate ion is equatorial.

The nomenclature of these conformations is derived from vague observations of the different shapes that the cyclic molecules can adopt, described by a simile. The shapes are important because they predetermine the interactions that are possible between the molecule and active site. For instance, the carboxyl group placed in an axial position on a chair conformation of sialic acid has a different directionality to the equatorial position that is observed for the pseudo-boat conformation, so the hydrogen bonding requirements for each conformation are different and this has consequences for enzyme-substrate binding. Consequently, sialic acid adopts the pseudo-boat conformation because this maximises hydrogen bonding in the ES complex, despite this being the energetically least favourable conformation. Other important binding regions include the interaction of a glycerol side-chain of sialic acid with glutamate residues and a water molecule by hydrogen bonding. The hydroxyl group at C4 is situated in another binding pocket and interacts with a glutamate residue, and lastly, the acetamido substituent fits into a hydrophobic pocket, which is important for molecular recognition. The pocket contains the hydrophobic residues Trp-178 and Ile-222 which lie close to the methyl carbon C11 and the hydrocarbon backbone of the glycerol side chain. Furthermore, the distorted pyranose ring binds to the floor of the active site with its hydrophobic face and glycosidic OH at C2 is moved into an axial position where it can form hydrogen bonds to Asp-151.

By understanding the binding of the substrate, chemists were able to propose a mechanism for the hydrolysis reaction. It was discovered to involve proton donation from an activated water molecule, facilitated by the negatively charged Asp-151 and formation of an endocyclic sialosyl cation TS intermediate. Then sialic acid is formed and released. This proposed mechanism was supported by kinetic isotope studies, which indicate an S_N1 nucleophilic substitution and NMR spectroscopy, a technique used to study the change in conformation of the substrate, showed that sialic acid is released as the α-anomer; consistent with an S_N1 mechanism. Also, site-directed mutagenesis showed that replacing the charged amino acids Arg-151 with lysine and Glu-227 by aspartate, so that stabilisation of the intermediate is repressed, activity of the enzyme is lost.

Once the structure of the TS and the mechanism of hydrolysis were thoroughly understood, companies could begin to develop inhibiters of the neuraminidase active site in an effort to generate antiviral agents against influenza. At the forefront of these drugs were Relenza and Tamiflu. The structures of these two drugs are compared in Figure 4.4.

Sialic acid analogues were synthesised with a double bond between C2 and C3 to replicate the trigonal planar geometry of the transition state of the hydrolysis reaction, based on molecular modelling programmes to evaluate binding interactions in the active site. This approach used the molecular fragments techniques, where probe atoms are situated in the binding pockets within the active site and the strength of the binding interactions are measured via energy calculations using software. The probe atoms constitute important functional groups, such as the carboxylate group, ammonium cation and hydroxyl groups, as well as the hydrophobic methyl group. Hydrogen bonding exhibits directionality, so the orientation of these fragments needs to be comprehended in order to attain the most favourable interaction energy.

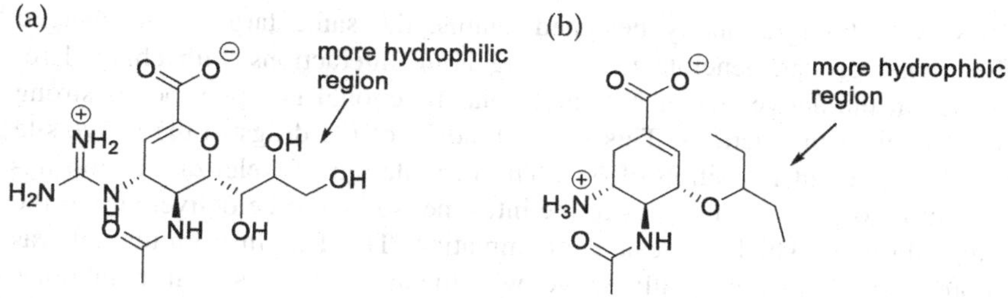

(a)   (b)

FIGURE 4.4 Molecular Structure of Anti-Influenza Drugs (a) Relenza™ and (b) Tamiflu™.

The next stage is to develop a molecular scaffold that places the functional groups in such a way as to optimise these interactions.

Once these parameters were ascertained, the relevant structures were synthesised and tested for activity. Molecular modelling predicted that the active site of the enzyme that normally binds the 4-hydroxyl group of sialoside is different in the viral enzyme compared to comparable bacterial and mammalian enzymes, which was confirmed by the crystal structure of inhibiter-bound enzymes, so selectivity could be achieved by replacing the 4-hydroxy with a different functionality; the larger guanidinium group was found to have greater hydrogen bonding interactions. This led to the structure of Relenza, developed by GSK, which was found to be a potent inhibiter of influenza NA. However, there were limitations regarding oral bioavailability (<5%) attributable to the polar nature of the molecule. Consequently, this drug has to be administered by inhalation.

Roche took a different approach to inhibition of neuraminidase. From studies of early sialic acid analogues, it was realised that the structure of the dihydropyran oxygen has no major role in binding to the active site; hence the possibility of replacing it with a methylene isostere to form a carbocyclic analogue; it would be advantageous to omit the polar oxygen atom, thus increasing the hydrophobicity of the molecule and potentially improve oral bioavailability. Furthermore, the glycerol side chain was removed to reduce polarity. This was replaced by a hydroxyl group to introduce an inductive electron-withdrawing effect on the carbocyclic double bond to alleviate its high electron density. Adding the hydroxyl group here facilitated the synthesis of ether groups on which hydrophobic substituents could bind to the pocket previously occupied by the glycerol side chain. As a result, a library of alkoxy analogues were screened and optimisation led to the development of Tamiflu, which could be taken orally.

Relanza and Tamiflu were developed separetely, but have similar structure and non-covalent interactions, which might be expected, given that they were both rationally designed against the same target. The charged functional groups generate very strong ionic interactions with charged residues in the active site of NA, and polar functional groups produce strong non-covalent interactions. This strong binding of the drugs to the active site results in potent inhibition of NA. The formulation of Relenza was too hydrophilic to cross membranes in the intestine, so had to be delivered into the lungs directly, which can cause inflammation. The Tamiflu formulation was more hydrophobic and orally active; with the ability to cross cell membranes more efficiently. However, Tamiflu needs to be taken within 24 hrs of flu starting to give the drug time to reach therapeutic levels in the lungs and give relief to symptoms.

The rational drug design of antiviral agents provides excellent examples of how medicinal chemists can strategically develop medicines against diseases by understanding the molecular basis of the disease and identifying appropriate targets. To combat viral infections, different strategies have been employed, with drugs developed against nucleic acid targets as well as protein targets. Rational drug design is not limited to producing therapies against pathogens: by understanding the biochemical processes that proceed within the human body, medicines can be made to treat derangements in the normal operation of the body, which lead to disease. Most notable perhaps is cancer research. By understanding the processes that lead to cancer, targets can be identified for drug design. Often, the targets for cancer therapy are on the DNA molecule. Indeed, many of the drugs targeting nucleic acids used in antiviral medicine originated from lead compounds for cancer therapy.

The Fight Against Cancer

5

CANCER: A COMPLEX DISEASE

Cancer is the second leading cause of death after heart disease; estimated to account for 9.6 million deaths globally in 2018, according to the World Health Organisation. It has become one of the most prevalent diseases in modern times, largely due to aging populations. Cancerous tissue develops from normal cells that have lost the normal regulatory mechanisms that control cell growth and multiplication. The abnormal cells multiply rapidly and uncontrollably to form cancerous tissue: neoplasm or tumours, which often lose their specialised characteristics that differentiate them from other types of cells. If the cancer is localised, it is said to be benign. If the cancer cells invade other parts of the body and produce secondary tumours, a process known as metastasis, the cancer is defined as malignant. This is the life-threatening form.

Cancer is termed a complex disease; there is no obvious single cause and there are more than 200 different types of cancer, which result from different cellular defects, therefore it is difficult to develop treatments to cure cancer. Therapies for one type of cancer may not be effective at treating another form of the disease.

Cancer can be caused by environmental influences, such as carcinogenic chemicals, example in cigarette smoke or in certain foods, which induce gene mutations or interfere with normal cell differentiation and division. Genetic mutation in genes that are responsible for controlling cell division can result in uncontrolled multiplication of cells to produce a tumour. In this way, mutagenic chemicals can result in carcinogenesis; the development of cancer.

The involvement of viruses in at least six human cancers has been observed, and are responsible for a significant proportion of the world's cancer deaths, though their involvement may not always be clear. Epstein-Barr virus causes Burkett's lymphoma and nasopharyngeal carcinoma; human papillomaviruses are STIs, which can lead to cancer in the cervix; and hepatitis B may be involved in many cancers of the liver. Viruses can cause cancer in several ways. They may bring oncogenes into the host cell and insert them into the genome, or some viruses may carry one or more promoters of gene mutation.

An individual may have a genetic predisposition for a certain cancer. Defective genes can be inherited, increasing the risk of cancer in subsequent generations. There are numerous possible genetic faults that can lead to cancer. Proto-oncogenes are genes that normally code for proteins that are involved in the control of cell division and differentiation. If they become mutated into an oncogene, this may disrupt the normal function of the gene and without control mechanisms in place, the cell could become cancerous. For example, the Ras protein is involved in the signalling pathway leading to cells division, during the processes of mitosis and meiosis. In normal cells, this protein has the self-regulating ability to switch itself off. In the mutated gene, Ras loses this ability, and is continually active leading to uncontrolled cell division. This mutation is found to be present in around one-fifth of human cancers.

In the event that DNA becomes damaged in a normal cell, there are recognition mechanisms in place so that the damage can be detected and DNA replication halted to give the cell time to repair the damaged DNA before the next cell division. If repair is not possible, the cell commits suicide; a process called apoptosis. Tumour suppression genes code for proteins that are involved in the processes of checking, repair, and apoptosis. If one of these genes, such as TP53, which codes for a protein with the same nomenclature, P53, becomes damaged, the P53 protein, which is crucial for the repair mechanisms, no longer functions and defects in the DNA persist; being carried to the next generation of cells. As the extent of the damage in the tissue increases, it becomes more likely that the tissue will develop into a carcinoma.

Genetic defects can result in numerous possible cell abnormalities; all of which are associated with cancer. These include: abnormal signalling pathways, such as insensitivity to growth-inhibitory signals and abnormalities in cell cycle regulation, evasion of programmed cell death (apoptosis), and limitless cell division, known as immortality. Cancerous tissue has the ability to develop new blood vessels; a process known as angiogenesis, which is required to nourish the fast-growing tissue. Tissue invasion and metastasis can occur, where cancerous cells spread to other parts of the body and can propagate cancerous tissue in other organs.

Most, if not all of these abnormalities, must manifest before a defective cell can develop into a terminal malignant growth, because a series of safeguards are in place to prevent adverse effects that may arise from a single cell defect. This explains why cancers can take many years to develop after prolonged exposure to harmful mutagens, such as asbestos or coal dust. Short-term exposure may cause a limited number of mutations, but cellular safeguards keep control over abnormal cellular activity. A lifetimes exposure to damaging mutagens, as with long-term tobacco smokers, who are frequently exposed to carcinogenic chemicals, so that the extent of damage that occurs overwhelms the safeguard mechanisms until the abnormal cell is liberated from the control mechanisms and becomes cancerous.

Cancers usually occur later in life as a result of exposure to mutagens over a longer timescale, and also because of age-related breakdown of control mechanisms. This is also one reason why cancer is difficult to treat: by the time cancer has appeared, several cellular defects have arisen and over-ridded multiple control mechanisms, so addressing one cell defect with a therapy is unlikely to be effective. Consequently, traditional anti-cancer agents tend to be highly toxic, acting on a variety of targets via different mechanisms; focussed on death of abnormal cells, but without high specificity over normal tissues. These drugs are referred to as being cytotoxic, so a dose rate must be carefully selected to ensure suppression of the carcinoma, but at the same time is bearable for the patient.

More recently, anti-cancer drugs have been developed that target specific abnormalities in cancer cells; improving selectivity and thus reducing side effects. However, cancer cells will contain multiple cells defects. Targeting one will not be an effective treatment; hence combination therapy is required, with drugs that target cancer via different mechanisms. This approach may be used in conjugation with radiotherapy or surgery.

Abnormal signalling pathways disrupt normal cell growth and division. The normal signalling pathway involves hormones, known as growth factors, which are extracellular chemical messengers which activate protein kinase receptors in the cell membrane. These receptors trigger a signal transduction pathway which eventually reaches the nucleus and promotes transcription of proteins and enzymes needed for cell growth and division. Defects in this signalling process results in the cell being constantly instructed to multiply; hence ultimately results in cancerous growths. The complexity of this signalling process means that there are several points of the cell cycle that can go wrong.

Furthermore, many cancer cells are capable of growing and dividing in the absence of external growth factors because they have the ability to produce their own growth factors; releasing them to stimulates the cells own receptors. Examples of growth factors include platelet-derived growth factor (PDGF) and transforming growth factor-α (TGF-α), which are important in the development of a carcinoma. Cancer cells may also produce abnormal receptors, which are constantly switched on, despite the absence of growth factors. Also, it is possible for a receptor to be over-expressed when an oncogene is too active; coding for excessive quantities of protein receptor in the cell membrane, which means that the cell becomes supersensitive to low levels of growth factor. For example, Ras protein receptors in cancerous cells can be over-abundant and usually lose the ability to auto regulate, being constantly switched on and the signalling pathway for cell division becomes over-active.

Additionally, there are several external hormones in place, such as transforming growth factor-β (TGF-β) that counteract the effects of stimulatory growth factors and signal for the inhibition of cell growth and division. Damage to the genes that code for the receptors of these inhibitory hormones, the tumour suppressor genes, can result in insensitivity to these signals and increase the risk of cells becoming cancerous.

Chromosomes containing mutated genes will become more prevalent if cell division and replication of the defective DNA is allowed to continue unimpeded, when the necessary control mechanisms fail, and this vastly increases the risk of cancer; as defective cells multiply in number, they can grow uncontrollably and defective genes can be inherited in gametes, if damage occurs during meiosis. Humans have 22 pairs of homologous chromosomes (X and Y) and a pair of sex chromosomes; a total of 46 chromosomes, known as the diploid number. The diploid number varies between species, for example dogs have 78. These are always even numbers because half comes from each parent. The chromosomes are dispersed throughout the nucleus and each comprise a single molecule of DNA that is supercoiled, held in place by protein molecules called histones. The chromosomes are only distinctly visible under the microscope during cell division.

Before cell division can occur, DNA replication must take place so that the genetic material can be passed to the daughter cells. DNA unwinds ready for replication, then the two strands are 'unzipped' by the enzyme DNA helicase, which breaks the hydrogen bonds between bases. The exposed bases of each strand can now undergo complimentary base pairing with free nucleotides in the nucleus, which are assembled by another enzyme, called DNA polymerase. Finally, all the nucleotides are combined to form two new polynucleotides; each comprised of an original strand and a new strand. This is called semi-conservative replication.

The daughter cells, produced after cell division, will now have the correct genetic information for controlling cellular processes. There are two forms of cell division: mitosis and meiosis. Mitosis produces two daughter cells and enables cells to replicate for growth and repair. Meiosis produces four daughter cells, each with half the number of chromosomes: haploid cells, which are important for making gametes, which fuse during fertilisation to make diploid cells, and ultimately a foetus.

Mitosis proceeds in five phases. During interphase, the cell is in its normal form; not dividing and the nucleus contains loose DNA, chromatin. •
During prophase, the nuclear envelope disappears and chromosomes become

visible. The chromosomes arrange themselves at the equator (centre) of the cell in metaphase, attached to spindle fibres. In anaphase, each of the two threads of a chromosome (chromatid) migrate to the opposite pole. By telophase, the nuclear envelope reforms and the cell splits to yield two identical diploid cells. If mutation occurs, the daughter cells will not be genetically identical to the parent cell and risk not functioning properly in the body.

Cells undergo a regular cycle of cell division, which is split into three periods. Interphase comprises the majority of the cell cycle, when there is no division, and is subdivided into a further three parts: first growth phase (G1), where organelles produce proteins, synthesis phase, when DNA is replicated, and second growth phase (G2), when organelles grow and divide and energy stores are increased. The second phase is nuclear division, where the nucleus divides into either two (mitosis) or four (meiosis). The last phase is cell division to produce the daughter cells. For mammals, the cell cycle takes about 24 hours, 90% of which is interphase. Most cancers are caused by damage to the genes that regulate mitosis and the cell cycle.

Abnormalities in cell cycle regulation may occur at any of the four major phases, known as G_1, S, G_2, and M. progression through these phases of the cell cycle depends on the balance of the chemical signals that promote growth or inhibition. The G_1 (gap 1) phase is where the cell grows in size and prepares for DNA replication in response to growth factors. The second phase (synthesis) is when DNA replication takes place. The next interval, once the chromosomes have been copied, is G_2 phase (gap 2) where the cell prepares for division. During this interval, the cell has time to check for errors in the DNA replication, and repair any damaged copies. The final phase, M (mitosis) is when cell division happens to produce two daughter cells, each containing a full set of chromosomes. The daughter cell then begins the cell cycle at the G_1 phase, or may remain in a dormant phase G_0.

During the cell cycle, there are various decision points to determine whether the cell should continue to the next phase or not. The restriction point, R, during the G_1 phase frequently becomes abnormal in cancer cells. Also, surveillance mechanisms are in place, known as check points, which assess the integrity of the process and a delay will occur in the G_2 phase if damage to DNA is detected. During this time, repair mechanisms can work to rectify the damage, or if the damage incurred is too great, the cell may commit suicide during apoptosis. If these checkpoints fail to operate properly, cancer can arise. The replication of malfunctioning cells, which lack the necessary mechanisms of quality control can lead to the development of a tumour because there is no mechanism in place to stop their uncontrolled division.

Among the mechanisms in place to control the cell cycle are cascades involving a variety of proteins called cyclins, along with enzymes called cyclin-dependent kinases (CDK). Different types of each are responsible for the regulation of the different phases of the cell cycle. Binding of a cyclin with its enzyme activates the enzyme and serves to move the cell cycle from one phase to the next. Progression through the cell cycle is regulated by sequential activation of cyclins and CDKs. The process can also be down-regulated by CDK inhibitors.

A number of restraining proteins are in place to regulate the effects of cyclins and have an inhibitory effect by blocking the activity of CDKs. The amount of restraining protein present is controlled by P53, which is important for monitoring the health of the cell and the integrity of DNA. Several cancers are associated with overactive cyclins or CDKs. This occurs when these molecules are produced in excess, or the genes that code for inhibitory proteins are malfunctioning. Indeed, half of all human tumours lack a properly functioning P53 protein, consequently the levels of restraining protein are low; in other words, there is insufficient production of CDK inhibitors and so the cell cycle proceeds unregulated.

The P53 protein also has an important role in apoptosis. Under normal circumstances, faulty cells that cannot be repaired have an intrinsic self-destruction mechanism, which is automatically initiated if the monitoring processes at checkpoints fail to recognise the necessary chemical signals that divulge corrupt DNA.

There are two distinct pathways for apoptosis. One: an extrinsic route, where three external factors govern whether or not apoptosis occurs; these are sustained lack of growth factors/hormones, T-lymphocytes of the immune system circulate the body and remove damaged cells, and death activator proteins, which may bind to tumour necrosis factor receptors on the cell membrane, triggering a signalling process that results in apoptosis. Two: an intrinsic pathway, where mechanisms within the cell detect damaged DNA and cause the production of more P53 tumour suppressor protein, which at sufficient levels triggers apoptosis.

Cancer cells can be described as 'immortal' because there is no apparent limit to the number of times that they can divide. Normal cells have a pre-determined lifetime, based on the number of times that their DNA can be copied, which is typically 50-60 cell divisions. Structures called telomeres have a key role in the immortalization process. The purpose of a telomere is to serve as a splice at the end of a chromosome and to stabilize and protect the DNA. It consists of a polynucleotide region at the 3' end of the chromosome consisting of several thousand repeats of a short (six base pairs) sequence. Around 50–100 base pairs are lost from the telomere after each replication because DNA polymerase is unable to replicate the 3' ends of the chromosomal DNA. Ultimately, the telomere becomes too short to function and the DNA becomes unstable and either unravels or links up end-to-end with another DNA molecule. This is a fatal error for the cell and apoptosis is triggered.

The immortality of cancer cells; their ability to divide an indefinite number of times, is consequential of the ability of cancer cells to maintain the length of their telomere by expressing an enzyme called telomerase. This is a type of RNA-dependent DNA polymerase, which works by adding hexanucleotide repeats on to the end of telomeric DNA, thereby maintaining its length. The gene encoding this enzyme is essential for the development of an embryo, but is supressed after birth. Cancer cells develop to remove this suppression and express the enzyme; achieving immortality as a consequence. It is found that telomerase is expressed in >85% of cancers.

Angiogenesis is the process by which new blood vessels are formed. As tumours grow, they require more essential nutrients, such as amino acids, nucleotides, carbohydrates, as well as oxygen for respiration, so the tumour must have a good blood supply. The larger a tumour becomes, its cells, particularly towards the centre of the mass, become more remote from the blood supply, so are starved of these resources. Tumour cells release growth factors to counteract this problem. Vascular endothelial growth factor (VEGF) and fibroblast growth factor (FGF-2) bind to cell receptors of the endothelial cells of nearby blood vessels and simulate the cells to divide, resulting in branching and extension of the existing capillary network to supply the ever-increasing demands of the tumour.

In normal cells, vascular growth factors are usually released when tissues have been damaged. Angiogenesis helps repair the injuries and is regulated by angiogenesis inhibitors, such as angiostatin and thrombospondin. Matrix metalloproteinases are enzymes that break down the membrane around the blood vessels to allow endothelial cells to migrate towards the tumour and enable the release of angiogenesis factors. Tumours are able to receive the blood supply they require due to a disruption to this balance. Furthermore, this increases the risk of cancer cells escaping from the primary source and metastasizing because of the increased availability of blood and newly developing endothelial cells can release proteins such as interleukin-6 that stimulates metastasis. The angiogenesis promoted by cancer cells result in the production of abnormal blood vessels, which are dilated, permeable, and have a disorganised structure. Also, integrin molecules are displayed on the cell surface membrane, which are usually absent from mature cells, so prevent apoptosis.

One possible way to combat cancer would be to inhibit angiogenesis. Drugs developed to breakdown abnormal blood vessels are likely to be less toxic than traditional chemotherapeutic agents, but need to be used along with other treatments. The angiogenesis inhibiters serve to 'normalise' the tumour capillaries before destroying them. This helps to gain access to the tumour with other anti-cancer agents, and stall cancer growth and shrink the tumour because fewer capillaries are present to supply the tumour with the essential nutrients to grow.

The selective delivery of anti-cancer agents can be achieved by taking advantage of the leaky capillaries produced by angiogenesis. Anti-cancer drugs encapsulated into liposomes or nanospheres cannot escape normal blood vessels, but can diffuse through the faulty ones in a tumour, so the anti-cancer drug accumulates in the tumour. Cancerous tissues tend not to develop an effective lymphatic system, so the polymeric drug delivery system becomes trapped at the tumour site, therefore when the drug is released, it is concentrated in the tumour.

In well-developed tumours, capillaries do not access all regions of the tissue, despite unregulated angiogenesis. These cells are deprived of oxygen and nutrients and may become dormant. This poses an issue for cancer therapy because most anti-cancer drugs work best on actively dividing cells. Even if a treatment has successfully eliminated the cancer, dormant cells start to multiply and the cancer reappears. Moreover, these cells are more likely to metastasize.

Tissue invasion and metastasis can result in cancer spreading to other parts of the body. Benign tumours are localised growths that remain in a particular part of the body, so are not life-threatening. By contrast, malignant cancer has the propensity for its cells to break away from the primary tumour and invade a blood vessel or the lymphatic system, be circulated, and propagate a tumour elsewhere in the body; ultimately with fatal results. To do this, cancer cells must overcome the constraints that are designed to keep cells in place.

To ensure that cells remain in the correct part of the body, they have a molecular signature on their surface with which they adhere to other cells of similar character and to the insoluble extracellular protein matrix, which holds the tissue together. Cells of the tissue are anchored to the extracellular matrix with the aid of integrin proteins. In the event that a normal cell becomes detached, apoptosis is triggered and this prevents cells from straying from one part of the body to another. Moreover, the adhesion molecules on the surface, the integrin proteins, must match the relevant extracellular matrix in order for the cell to adhere and survive.

When cancer cells metastasise, the integrin molecules are absent. Inevitably, they detach from the primary tumour, yet are able to survive independently and can attach to the extracellular matrix in another part of the body. Apoptosis is not triggered and the immigrant cells can propagate a secondary tumour. It is hypothesised that the oncogenes code for false protein messages that imply that the cell is still attached.

Cancer cells and white blood cells are the only types of cells that are able to breach the extracellular matrix; both possess the metalloproteinase enzyme that hydrolyses the proteins that comprise this barrier. Once this barrier has been breached, cancer cells are free to penetrate blood vessels and enter the blood supply. The cancer cells then circulate in the blood and usually become trapped in the first network of capillaries that they encounter; for most tissues, secondary tumours establish in the lungs, intestines, or liver.

INTRODUCTION TO CANCER CHEMOTHERAPY

Among the doctors arsenal for the treatment of cancer is surgery, radiotherapy, and chemotherapy. Often a combination of these therapies is implemented, but for the medicinal chemist, chemotherapy is the pinnacle of cancer research. A variety of anti-cancer drugs may be used in combination, having different mechanisms of action in order to improve the likelihood of a successful treatment. Traditional anti-cancer treatments often affect normal cells as well as cancer cells due to their similarity, which makes it difficult to identify discrete targets, hence the renowned toxicity of chemotherapy. The challenge for the medicinal chemist is to develop a drug that will work on a specific target, or at least to devise a method of drug delivery that ensures the drug becomes concentrated in the tumour, thus minimising side effects.

Due to the nature of cancer, cells rapidly dividing, fast growth, the large requirement of nutrients, and the large network of capillaries to supply them means that drugs do naturally tend to accumulate in cancerous tissue. However, a disadvantage is that dormant cells can evade the drugs, while normal cells that divide rapidly, such as bone marrow cells, can accede to the effects of the drug. This can lead to a weakening of the immune system because the chemotherapy impacts on the production of white blood cells. The patient therefore has less resistance to infection, making the individual susceptible to pathogens and prone to infections.

Most traditional anti-cancer agents are cytotoxic; they work by disrupting the function of DNA in order to destroy abnormal cells. Some act directly on DNA while others, known as antimetabolites, affect the enzymes involved in DNA replication. Cancer research has led to a better understanding of the molecular mechanisms of abnormal cell chemistry. Knowledge of the cell cycle and genetic analysis have aided the development of new treatments, such as less toxic forms of chemotherapy, the use of monoclonal antibodies, and enzyme pro-drug therapy. Progression in anti-cancer research is now directing towards anti-cancer agents that act on specific molecular targets that are presented by cancer cells.

A challenge for cancer treatment is the occurrence of resistance, which can be intrinsic or acquired. Intrinsic resistance refers to a scenario where the anti-cancer drug is not effective at any stage of the treatment. This could be due to biochemical or genetic reasons, slow growth of the tissue, or poor uptake of the drug, for example by dormant cells. Acquired resistance is where the drug initially is effective against the tumour, but the cancer cells develop to lose their susceptibility to the drug. This often results from the destruction of drug-sensitive cells to enrich the tumour in resistant cells present in the tumour, which continue to replicate and multiply. Ultimately a tumour grows that does not respond to the chemotherapy. This property of the tumour derives from genetic mutation of some cancer cells attributable to their genetically unstable disposition.

Resistance can arise through several molecular mechanisms that stem from mutation. Notably, the predominant avenues for tumour resistance include: decrease uptake of the drug, for example by dormant cells, increased synthesis of a given target molecule beyond the scope of the therapeutic index of the drug; changes in metabolic pathways meaning that the drug is no longer metabolised and therefore may no longer activate the pro-drug. Furthermore, efflux pumps, p-glycoprotein transport channels, can expel a range of molecules and can result in multi-drug resistance, hence efforts to make p-glycoprotein inhibitors are an active area in cancer research.

Different strategies can be employed to target DNA; producing drugs that work by different mechanisms to help regress the advancement of cancer. Intercalating agents and alkylating agents target DNA directly by distorting the shape of the helix and cross linking the two nucleic acid strands, respectively. This has the result of interrupting DNA replication and supresses cell multiplication. Alternative strategies involve targeting the enzymes involved in DNA synthesis; hormone therapy antagonises the receptors that promote DNA replication, and structural proteins that are important for mitosis can also be targeted.

Certain natural products extracted from microbes were found to have an impact on the rate of cancer growth. These compounds acted as intercalating agents. The molecule inserts into the grooves along the sugar-phosphate backbone and binds by non-covalent means causing a distortion to the shape of the helix. This has the effect of preventing the enzymes involved in DNA replication and transcription from operating on the molecule. Intercalators are flat aromatic compounds, which can fit into the grooves of the DNA helix, where heteroatoms form hydrogen bonds between the base pairs.

An example of a naturally occurring intercalating agent is doxorubicin; isolated from *Streptomyces peucetius*; it belongs to a group of antibiotics called anthracyclines. It has a tetracyclic ring system, where three of the rings are planar and intercalate the DNA double helix in the major groove. The charged amino group on the sugar ring forms an important ionic bond with the negatively charged phosphate groups of the DNA backbone, crucial for the drugs activity.

The mechanism of activity against DNA is intercalation-mediated topoisomerase inhibition, which prevents the normal function of the topoisomerase II enzyme. As the DNA double helix unwinds during the replication process, increased tension and entanglement can occur further down the helix. Where two sections of DNA are in close contact, the enzyme will bind to one of the double helices and a tyrosine residue in the active site is used to score both strands of the DNA by attacking a phosphate and cleaving the DNA backbone. This produces a temporary covalent bond between the 5' end of the stand and the nucleophilic tyrosine oxygen of the enzyme, thereby stabilising the DNA. The intact section of DNA can now pass over the other strand, avoiding entanglement. The enzyme then reconnects the strands and departs.

With the intercalation of doxorubicin, the DNA-enzyme complex is stabilised such that the covalent bond to tyrosine is not cleaved and the enzyme cannot depart. This can lead to numerous DNA breaks, which, when encountered by DNA polymerases and helicases during the replication process, may trigger apoptosis. This is effective against rapidly proliferating cancer cells, where topoisomerase II is very active.

Mitoxantrone is a synthetic analogue to the anthracyclines, with a similar mechanism of action, used in the treatment of leukaemia and lymphomas predominantly. Its structure is simplified to allow for an uncomplicated synthesis, and lacks the sugar moiety of the anthracyclines, which is thought to cause cardio-toxic side-effects. However, the amino group of the sugar moiety is important for binding, so needed to be replaced by a suitable substituent that would place the required nitrogen in the same relative position. The tetracyclic ring system was truncated by removing the non-planar ring and replacing it with two identical substituents to make the molecule symmetrical and therefore easier to synthesise. The planar tricyclic ring system intercalates the DNA in much the same way as the anthracyclines, with studies of structure-activity relationships revealing the carbonyl and phenol groups to be an important pharmacophore, while the amine nitrogen is crucial for binding. The structures of the anthracycline drug doxorubicin and synthetic analogue mitoxantrone are given in Figure 5 of the Supporting Material*.

Mustard gases were used in the First World War to devastating effect. It was later realised, however, that nitrogen mustards had potential as anti-cancer agents. They are classified as alkylating agents, referring to their mechanism of action. These compounds are highly electrophilic and readily react with nucleophiles to form strong covalent bonds. Alkylating molecules will react with the nucleophilic nitrogen groups of DNA bases, particularly N-7 of guanine. Symmetrical molecules are used, with electrophilic groups on either side, which will react with guanine bases on each strand of DNA and therefore cross-link them, which causes transcription and translation to be disrupted and therefore impede the progress of cancer. Conversely, the drug may link to two guanine bases on the same strand of DNA and cover that portion of the DNA from the actions of enzymes involved in replication. Predictably, there are limitations to alkylating drugs. Such reactive compounds will attack other

---

* Available at www.routledge.com/9780367644031

nucleophilic groups on proteins also, so have poor selectivity, and themselves are mutagenic due to their action on DNA.

The first alkylating agent to be used medicinally was chlormethine, in 1942. Its mechanism of action, shown in Figure 5.1, involves displacement of a chloride ion via intramolecular nucleophilic substitution to produce a highly electrophilic aziridium ion. Alkylation of DNA can then occur when the nucleophilic N-7 of guanine attacks the β-carbon of the electrophile. This process is repeated with the other strand of DNA to cause cross-linking, halt DNA replication, and thus serve as an anti-cancer agent.

The nitrogen mustards are highly reactive and will react with other nucleophiles besides guanine. They are far too reactive to survive the oral route, so must be administered intravenously. In order to modulate activity and reduce unwanted side reactions, electron-withdrawing substituents can be added to the nitrogen atoms in place of Me, such as an aromatic ring. This has the effect of making the lone pair of electrons on the nitrogen less available for displacing the chloride ion and the rate of formation of the aziridium ion is reduced. In this example, the lone pair interacts with the resonating electrons of the aromatic ring, so only strong nucleophiles like N-7 of guanine are able to force the alkylation reaction.

Strategies can be adopted to improve the selectivity of alkylating agents. Attaching a moiety to the aromatic ring to mimic the structure of phenyl alanine, so that the drug is recognised as an amino acid, allows the drug to be readily taken into cancer cells, where the demand for amino acids is greater, via transport proteins. A similar approach is used with uracil mustards, where the nucleic acid building block uracil is attached to the mustard. Again, selectivity is displayed as a consequence of the greater demand for nucleic acid bases by tumours compared to normal tissue, so higher concentrations of the drug accumulate in the tumour.

Metal complexes, such as those illustrated in Figure 5.2, have been used as cross-linking drugs to target DNA directly. Cisplatin is frequently used in the IV treatment of testicular and ovarian cancer. Its discovery was fortuitous from research in the 1960s on the effects of electric currents on bacterial growth. It was found that bacterial growth was inhibited by an electrolysis product from the platinum electrode: cisplatin.

Cisplatin acts primarily as a DNA 1, 2-intrastrand cross-linker, and its molecular geometry is key to its chemical mode of action. The molecular geometry, which is flat, referred to as square planar, arises from the nature of the bonding that is present between metals and their ligands. For the organic molecules considered so far in this text, covalent bonding involved shared pairs of electrons spaced at the furthest distance possible to yield a tetrahedral geometry, with bond angles of 109.5 degrees, as predicted by the VSEPR model. In the case of metal-ligand bonding, both the electrons are supplied by

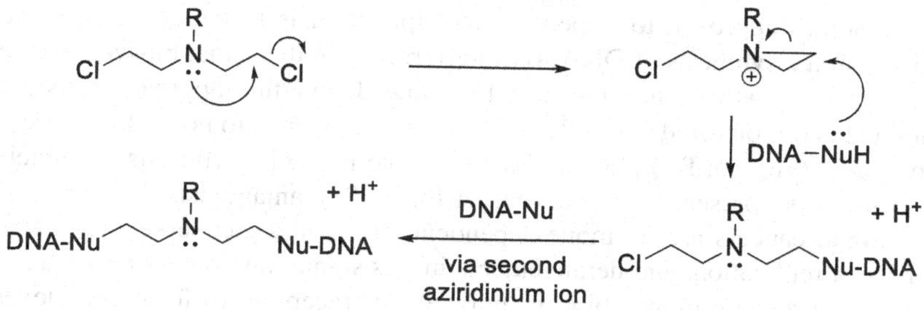

FIGURE 5.1 Mechanism for the Cross Coupling of DNA by Nitrogen Mustards.

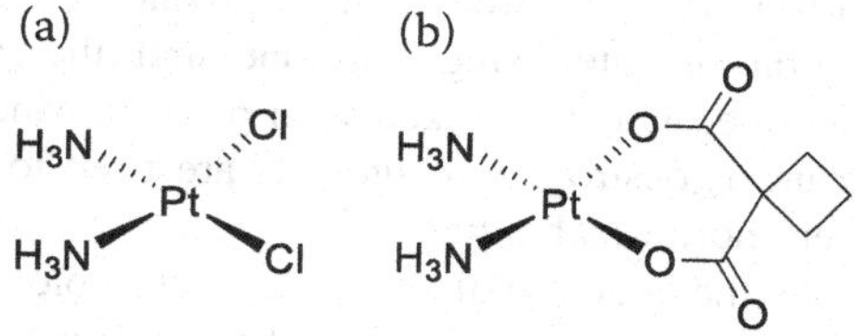

FIGURE 5.2 Structure of the Platinum Complex Known as Cisplatin (a) and the Second Generation Analogue Carboplatin (b).

the ligand in what is referred to as a coordinate bond. This occurs because metals have a different valence electron configuration to other types of atom, with defined energy levels that allow for this type of bonding to predominate. In the case of cisplatin, the ammonia and chloride ligands each donate their lone pair of electrons to the platinum ion, and the occupancy of these electron shells dictates the square planar geometry, with bond angles of 90 degrees. The geometry of the metal complex is key to the mode of action.

The chloride ligands are displaced by a water molecule *in vivo*, and the $[Pt(NH_3)_2(H_2O)]^+$ complex approaches the negatively charged DNA. The water ligand is replaced when Pt forms a coordinate bond to N7 of the nucleobase guanine, causing distortion of the purines and the DNA helix becomes kinked. As a result, DNA replication and repair cannot occur and the cell dies. Cisplatin undergoes this process very aggressively, and consequently leads to severe side effects. Attempts to produce analogues that are less toxic involved the utilisation of organic ligands. The organic molecules in place of the two chlorides have two donor groups, so two coordinate bonds are formed with one ligand, making the ligands more difficult to displace by a water molecule, hence the activity is reduced.

Another approach to impede DNA replication is to target enzymes involved in the synthesis of DNA. By interfering with the actual biosynthesis of DNA, the growth of tumours can be reduced, but this approach must be targeted and monitored carefully to avoid undue damage to normal processes. Folic acid (vitamin B9) plays an essential role in the biosynthesis of nucleic acid bases, so presents a prudent target for finding antagonists.

Some cancers are hormone-dependent. Hormones can serve as a signal for DNA replication, and derangements in this signalling process may lead to cancer. Steroid hormones bind to intracellular receptors to form complexes that act as nuclear transcription factors to control whether transcription takes place. Hormone antagonists can be used to block these receptors, thus impede cell proliferation. Oestrogen is commonly involved in these pathways. Oestrogen drugs are used in the treatment of prostate cancer because they inhibit the production of luteinising hormone and thereby decrease the synthesis of testosterone, which is linked with prostate cancer. In hormone-dependent breast cancer, oestrogen antagonists are used to block oestrogen receptors and prevent hormone binding.

Look how closely the structure of stilboestrol resembles that of oestrogen (Figure 5.3) if bonds are imagined between the ethyl groups and the 3-position of the phenol rings, therefore it is unsurprising that stilboestrol acts as an agonist in the oestrogen receptor. The structure of tamoxifen is also similar, to the extent that it can bind to the oestrogen receptor, but is sufficiently different not to initiate a response upon binding, therefore acts as an antagonist. Note that the E stereochemistry of the alkene is important in stilboestrol for binding, and indeed the E isomer of tamoxifen acts as an oestrogen agonist also. The OH group is also important for binding. The two hydroxyl groups are separated by what is essentially a hydrophobic spacer so that they are in the correct position relative to each other to fit specifically into the oestrogen receptor site. The two hydroxyl groups nestle into polar pockets within the receptor; crucially the phenolic group forms hydrogen bonds with three groups. Molecular shape complementarity arises as the cyclic middle fits into a cavity formed by hydrophobic amino acids.

The binding cavity of the oestrogen receptor is both rigid and flexible. Rigid regions mediate recognition of the phenolic component of the ligand, as well as the separation of the hydroxyl groups. Flexible regions accommodate extension of the ligand core, for example inclusion of antagonistic side chains, and permits alternate binding modes of the spacer component. Refer to Figure 5.4. Remember that drug efficacy relates to biological activity, not strength of ligand binding. The structural basis of agonism/antagonism relates to competition for the surface responsible for co-regulators binding: different ligand classes induce or stabilise the distinct orientation of helix-12; a region of the oestrogen receptor that dictates the formation of the co-regulator binding cleft. In this way, oestrogen

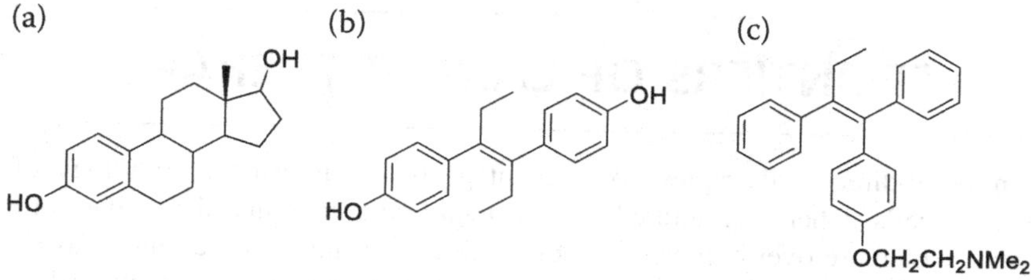

(a)   (b)   (c)

FIGURE 5.3 Molecular Structures of Oestrogen (a), Stilboestrol (b), and Tamoxifen (c).

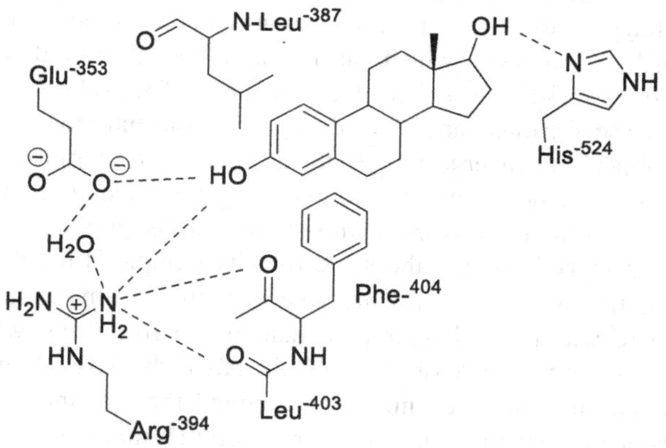

FIGURE 5.4 The Binding of Oestrogen with Key Residues in the Oestrogen Receptor.

analogues can be designed with different side chains to achieve different biological activity. This illustrates how rational drug design can be adopted to produce novel therapies to battle diseases; in this case, cancer.

Another strategy to battle cancer is to target the structural elements of the cell division process. Disruption of mitosis will obviously impede cell proliferation. Tubulin is a structural protein that is crucial for cell division, involved in polymerisation and depolymerisation of microtubules, where it is used as protein building blocks. As the cells divide, mitotic spindle fibres link the two daughter cells. The spindle fibres are made from microtubule polymers formed from tubulin proteins. Drugs designed to bind to tubulin prevent its polymerisation to from the microtubules, and thus cell division is arrested.

FRONTIERS OF CANCER THERAPY

Antibody-directed therapies take advantage of the unusual morphology of cancer cells, where the altered plasma membrane contains distinctive antigens, which are over expressed. These antigens are much more numerous on cancer cells than normal cells and monoclonal antibodies can be utilised as anti-cancer agents, with the desired mode of action being to activate the body's immune system to direct killer cells towards the tumour. The antibody can also act as a receptor antagonist.

The activity of monoclonal antibodies in this instance is quite low. An improved strategy is to attach an anti-cancer drug to the antibody in the form of an antibody-drug conjugate, which can facilitate the selective delivery of the anti-cancer drug to specific types of cancer cells. This selectivity is highly desirable because, for many anti-cancer drugs, the concentration needed for the effective treatment of cancer is close to the threshold concentration for toxicity.

The first-generation antibody-drug conjugates used anti-cancer drugs such as methotrexate and doxorubicin, but yielded disappointing results with lower anti-cancer activity and the same toxicity compared to administration without the antibody as a vector. This is because the lifetime of the antibody-drug conjugate was much longer than that of the free drug, which was a contributing factor to the toxicity problem. Also, the size of the antibody impeded penetration into the tumour and limited the amount of drug being delivered, hence reduced activity. It was realised that more potent anti-cancer drugs were needed for application in antibody directed therapy. These drugs would require a stable bond to the antibody, so that it remains bound until reaching the tumour and not release the drug into the blood stream, which would result in high toxicity.

A drug can be bound to an antibody in a number of ways. Lysine residues are present in many parts of the antibody molecule and contain a nucleophilic primary amine functional group on to which a drug molecule could be added by simple acylation or alkylation reaction. However, care must be taken not to attach the drug to groups present in the binding region, or recognition site, for the cancer antigen. The drug must not obstruct antigen-antibody binding. To avoid obscuring the antibody binding surface, a good approach is to reduce the four intrastrand disulphide bridges at the hinge of the Y-shaped antibody protein molecule to produce eight thiol groups to which a drug may be added by alkylation. While this approach limits a maximum of only eight drugs being attached, linker molecule can be attached on to the antibody, which can contain several drug molecules. Alternatively, the carbohydrate region of an antibody can be utilised for drug attachment by lightly oxidising the vicinal diols of the sugar rings to produce aldehyde groups on to which drug molecules can be linked.

This linker must be cleaved only when the drug-antibody conjugate has entered the cancer cell, otherwise the method will lose its effectiveness. For example, a disulphide linker may be cleaved by disulphide exchange with an intramolecular thiol, such as glutathione, which has a higher concentration inside the cancer cell than in plasma. It is crucial that the drug is only cleaved inside the cell because drugs used in this application are highly potent, with IC_{50} values $<10^{-10}$ mol dm^{-3}, which are 100–1000 times more cytotoxic than conventional anti-cancer drugs.

Antibody-directed enzyme prodrug therapy (ADEPT) is another method utilizing the selectivity of antibodies. The mechanism for this approach is shown in Figure 5.5. In this instance, an antibody-enzyme complex is administered, where the antibody, which is linked to a bacterial enzyme, binds selectively to tumour antigens, but remains attached to the surface of the cell and must not be internalised, in contrast to antibody-drug conjugate therapy. After sufficient time has lapsed for all the antibody-enzyme complex to become bound to its target antigen, or expelled from the blood stream, a cytotoxic prodrug is administered. The design of these drugs ensures stability in the blood stream and they can only be cleaved to reveal the active compound by the enzyme complexed to the antibody, thus ensuring that the highly cytotoxic drug is concentrated in the tumour and the effect on normal cells is minimal.

Cleavage of the active cytotoxic drug occurs by enzymatic breakdown of the β-lactam moiety. A nucleophilic group in the active site attacks the unstable β-lactam at the electrophilic carbonyl carbon resulting in ring-opening, while the electrons from the bond resonate along the molecular structure to cause two further bond cleavages and release the active compound. Because the enzyme is anchored to the antibody, which selectively binds to cancer cells, the prodrug only encounters the enzyme within the cancerous tissue, so this is the only location where the active compound is unveiled, thus minimising side-effects.

This method has advantages over antibody-drug therapy. Enzymes are catalytic, so can generate a large number of active drug molecules at the tumour site, which diffuse into the tumour and may also target cancer cells in this region that do not display the necessary antigens for antibody binding. The use of foreign enzymes derived from bacteria circumvents the problems that would arise if a mammalian enzyme was to be used, which could result in similar enzymes within the body activating the highly cytotoxic prodrug in healthy tissue.

Complications could arise from the use of foreign enzymes in ADEPT in regards to immune responses. The use of human enzymes, as mentioned earlier, carries the risk of the prodrug being activated elsewhere in the body.

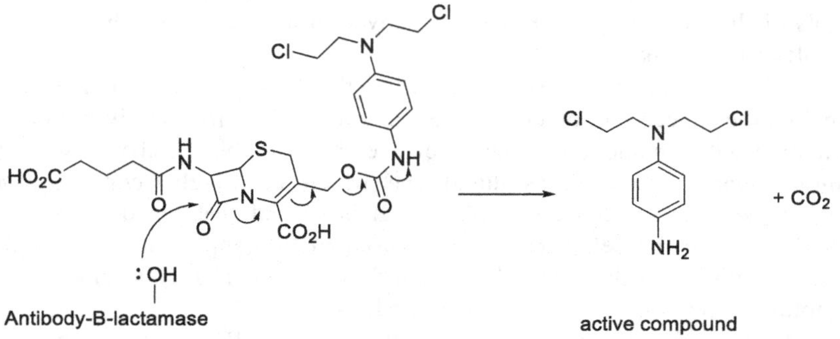

FIGURE 5.5 Mechanism for the Enzymatic Cleavage of a Pro-drug to Produce the Active Compound for the ADEPT Strategy to Tackle Cancer.

Also, the activity of human enzymes may be lower than bacterial enzymes, so lower levels of drug would be present in the tumour. However, research into gene therapy has produced techniques that could rectify these problems.

Gene-directed enzyme prodrug therapy (GDEPT) uses implanted genes into the cancer cell's DNA, which codes for the enzyme used to activate the prodrug. This is achieved through use of a vector, such as a retrovirus or adenovirus, where the desired gene is spliced into the viral DNA, and so is inserted into the host cancer cell on infection. Additionally, the virus is modified so that it is no longer virulent and cannot harm normal cells. Typically, the vector is administered directly into the tumour for greatest effect.

Once the necessary gene has been inserted, the cell will start to produce the desired enzyme. Because the foreign enzyme is produced inside the cell, it will not be subjected to immune responses, so this technique circumvents this issue that had important consequences for ADEPT. However, it is unlikely that the desired gene will be delivered to all the cancer cells, so it is important that the drug can be transferred between the cancer cells.

Further gene-directed research is at the forefront of modern cancer research. By understanding the molecular basis of this complex disease unique and unequivocal targets can be identified for which treatments can be developed that have much lower side-effects compared to the traditional anti-cancer therapies. Cancer is one of the main challenges to modern medicine, where prevalence among an aging population has increased, and the consequences of the disease are devastating. The pioneering research done by medicinal chemists though offers a chance for us to overcome this disease and perhaps create a future where the threat of cancer in old age is less ominous.

Targeting the Nervous System

6

THE PERIPHERAL NERVOUS SYSTEM

The nervous system is composed of two parts: the central nervous system (CNS; brain and spinal cord) and peripheral nervous system, which extends the entire body. Sensory neurons carry information from the body to the CNS, while motor nerves carry messages from the CNS to the rest of the body. Information from a stimulus is carried to the CNS by the sensory neurones; here, the messages are coordinated and the appropriate messages are then sent from the CNS to effectors, which can be organs or muscles, to generate a response to the stimulus. For example, if a person was to touch a hot object, for example they accidentally touch the ring on a hot cooker, the information from the stimulus, measured by temperature receptors in the skin, travels via a sensory neurone to a coordinator, such as a connector neurone within the CNS, which produces an automatic response to move the hand away from the heat by sending signals down motor neurones to the effector muscles. This pathway of neurones is known as a reflex arc. The response is immediate because only three neurones are involved and it is an automatic, involuntary response because it bypasses the brain to save time and avoid damage.

These signals can be considered as electrical pulses, but rely on the movement of ions, rather than electrons. As the ions move across the membrane of a neurone (nerve cell), a potential difference is created, which can be reversed to propagate an electrical signal; a nervous transmission, known as an action potential. Nerves are composed of many tightly packed neurons forming a bundle held together by connective tissue, and have a blood supply. They are distributed throughout the body like cables, and run in distinct pathways. A nerve may contain both sensory and motor neurones, but the proportions of each can vary.

A nerve impulse is a self-propagating wave of electrical disturbance, not current, that travels along the axon membrane of a neurone, due to a temporary reversal of electrical potential difference. This reversal alternates between two states: the resting potential and an action potential. There is a threshold level of stimulus needed to propagate an action potential; this feature is referred to as the 'all or nothing principle'. The magnitude of a stimulus is perceived by the number of impulses passing in a given time and/or having neurones with different threshold values, which is left for the brain to interpret.

A resting potential of around -65 mV is maintained by the active transport of Na^+ ions out of the nerve axon and K^+ ions into the axon via intrinsic protein channels that make up the 'sodium-potassium pump'. Three sodium ions are pumped out for every two potassium ions that enter the axon. Meanwhile, K^+ ions are able to diffuse back out of the axon through specific ion channels, while sodium ion channels are mostly closed. This creates an electrochemical gradient; tissue fluid outside the neurone is positive compared to the axon membrane, which is said to be polarised.

The energy from a stimulus causes a temporary reversal of charge in the axon membrane, due to the opening of voltage-gated channels. As the action potential passes, the sodium-potassium pump is turned off, while sodium voltage-gated channels open, allowing Na^+ to diffuse into the axon along their electrochemical gradient and potassium ion channels close. This causes a reversal of charge across the axon membrane and an action potential of $+40$ mV is generated. This wave of depolarisation proceeds along regions of the axon, causing sodium ion channels to open, and then close in a Mexican wave-like fashion. These channels then remain closed during a refractory period to ensure that the nerve impulses are discrete and unidirectional. Repolarisation occurs when the potassium ion channels open again and K^+ diffuse back out of the axon and the sodium-potassium pump reactivates.

A nerve route may be comprised of several different neurones. The point at which two neurones connect is called a synapse. Synapses act as junctions when nerve impulses are being transmitted, allowing a single impulse along one neurone to be transmitted along a number of different neurones at a synapse, allowing a single stimulus to generate a number of simultaneous responses. Equally a number of impulses can be combined at a synapse, which allows one response to be generated for stimuli from different receptors.

Where two neurones meet at a synapse, the nerve impulse is interrupted because there is a gap between the neurones, called the synaptic cleft. In order for the signal travelling down the presynaptic neurone to be continued, it must be propagated in the postsynaptic neurone across the synapse via neurotransmitters, such as acetylcholine. Neurotransmitters are molecules that, upon activation by an action potential, are released from the presynaptic neurone into the synaptic cleft. The neurotransmitter molecules diffuse across the gap and bind to receptors on the postsynaptic neurone, causing a chain of reactions that result in the propagation of the action potential in the post-synaptic nerve and the signal can proceed to its final destination. The actions of the neurotransmitter acetylcholine are instrumental in the operation of the cholinergic nervous system.

Acetylcholine is synthesised from choline and acetyl coenzyme A in the nerve ending of the presynaptic nerve, catalysed by the enzyme choline acetyltransferase. Acetylcholine is incorporated into membrane-bound vesicles by means of specific carrier proteins. An action potential results in the opening of calcium ion channels and increases the intracellular Ca^{2+} concentration, which induces the vesicles to fuse with the cell membrane of the presynaptic nerve and release acetylcholine in to the synaptic cleft. The neurotransmitter crosses the synaptic cleft and binds to cholinergic receptors on the surface of the post-synaptic neurone. This has the effect of opening sodium ion channels in the post synaptic membrane and propagating the action potential in the post synaptic nerve; whereby the signal can carry on to its destination. The enzyme acetylcholinesterase hydrolyses acetylcholine into choline and ethanoic acid, which diffuse back into the presynaptic neurone. This breakdown prevents new action potentials from being continuously generated.

Nerves transmit signals between the CNS, made up of the brain and spinal cord, to the body (peripheral nervous system). The motor nerves of the peripheral nervous system are divided into three subsystems. The somatic motor nervous system carries signals from the CNS to skeletal muscle to stimulate voluntary muscle contraction. The autonomic nervous system carries messages from the CNS to smooth muscle, cardiac muscle, and the adrenal medulla to stimulate the release of adrenaline. These nervous impulses are divided across two pathways. Parasympathetic nerves leave the CNS, travel some distance before encountering a synapse with a second nerve, and then the nerve impulse is transmitted across the junction using the neurotransmitter acetylcholine. Sympathetic nerves leave the CNS, but almost immediately synapse with a second nerve, again using acetylcholine. The second nerve proceeds to the same target organs as the parasympathetic pathway, but the synapses also have different receptors, that use different neurotransmitter: noradrenaline. These two systems tend to have an antagonistic relationship, where the sympathetic nervous system gets the body ready for action, and the parasympathetic calms the body down.

The enteric nervous system is located in the walls of the intestine and receives messages from parasympathetic and sympathetic nerves as well as responding to local reflex pathways, involving a variety of neurotransmitters.

The enteric nervous system tends to act more independently and is occupied with habitual modulation of the gastrointestinal system.

The sympathetic system promotes the 'flight or fight' response, where noradrenaline is release and promotes contraction of cardiac muscle, associated with increased heart rate, and relaxes smooth muscle, reducing contraction of GI tract and urinary tract. The elevated heart rate and suppression of general bodily operations are in readiness for action. Moreover, stimulation of the adrenal medulla releases the hormone adrenaline, which reinforces the process. The parasympathetic pathway leads to opposite effects: acetylcholine is released to target organs and stimulates bodily processes.

These opposing systems are kept in balance in the body, where the proportionate strength of each response depends on physical stimuli. It is crucial for a body to remain in homeostasis in order to function properly; failure in either of these systems could lead to ailment in the heart, skeletal muscle, or digestion, which could be a consequence of deficiency or excess of neurotransmitter. Hence, to rectify these ailments, there is the potential for developing drugs which act as agonists/antagonists to serve as treatments for conditions that affect these organs.

Consider the processes taking place in the cholinergic signalling system, as messages are transmitted between neurones or from a neurone to an organ, such as a muscle, across a synapse, using the neurotransmitter acetylcholine. The neurotransmitter is released from the presynaptic neurone during an action potential. The acetylcholine molecules travel across the synaptic cleft and must bind to cholinergic receptors in the membrane of the postsynaptic neurone. This then leads to processes that propagates an action potential in this nerve. The cholinergic receptor therefore presents an attractive target for drugs that can be designed to influence the nervous system.

The purpose of cholinergic receptors (autoreceptors) present at the terminal of the presynaptic nerve is to provide a means of local control over nerve transmission. The binding of acetylcholine to the autoreceptor has the effect of inhibiting further release of acetylcholine. Also, the presynaptic nerve contains noradrenaline, which serves as another system of presynaptic control of acetylcholine release. Noradrenaline is released when the sympathetic nervous system is active and binds to those receptors in the cholinergic synapse and has the effect of inhibiting the release of acetylcholine. If derangements occur, where there is an insufficient release of acetylcholine, it would be logical for medicinal chemists to develop an agonist for the cholinergic receptor. While acetylcholine can be readily synthesised in the laboratory, it is not feasible as a treatment due to being hydrolysed in the stomach and blood and there is no selectivity of action. Hence, analogues of acetylcholine must be developed that are stable to hydrolysis and selective where they act in the body.

Selectivity can be achieved through designing synthetic analogues with subtle structural differences that result in optimal binding with a particular cholinergic receptor at one part of the body over another. This is possible as slight differences in the structures of these receptors is observed at different parts of the body. Also, the drug might be preferentially distributed to one part of the body over another. There are subtle differenced in cholinergic receptors around the body, for example although the binding region for acetylcholine is conserved, there may be a molecular barrier present on the periphery that might block access to a larger synthetic analogue. This is also true for other types of receptors.

Differentiation among cholinergic receptors was first observed from studies into the physiological effect of nicotine (present in tobacco) and muscarine (the toxic compound in poisonous mushrooms). These are both acetylcholine agonists, but nicotine was found to be active between different nerves at the synapses between skeletal muscle and its motor units, but had poor activity elsewhere. Muscarine, by contrast, had poor activity in these places, and was found to affect synapses of nerves with smooth muscle and cardiac muscle. It was concluded that these two types of receptor, nicotinic and muscarinic receptors, had structural differences and demonstrated that receptor selectivity was possible. However, now there was the challenge to develop drugs without side effects.

This led to the study of structure-activity relationships (SAR) of acetylcholine to understand receptor binding and elucidate which aspects of the compound could be modified for selectivity. The positively charged nitrogen atom is essential, as is the distance from the atom to the ester group, so the overall size of the molecule cannot be altered significantly and the ester bridge must, therefore, remain intact. Once the essential features of the acetylcholine-receptor complex were identified and understood, medicinal chemists could work to develop compounds that would successfully bind with a particular cholinergic receptor, but have subtle structural differences that would give these compounds receptor selectivity.

The conclusions drawn from this research is that acetylcholine fits tightly in its binding site, so there is little opportunity for variation. Important hydrogen-bonding interactions presumably exist between the ester group of acetylcholine and an asparagine residue, while a small hydrophobic pocket is present into which the methyl ester can fit, but nothing larger. From the evidence, it appears that the NMe_3^+ group is placed in a hydrophobic pocket lined with aromatic amino acids made of three compartments; two of which are only large enough to accommodate methyl groups, while the third has enough space to tolerate a larger substituent. It is proposed that a strong ionic interaction exists between the positive nitrogen and an anionic aspartic acid residue. An alternative suggestion involves an induced dipole interaction between the NMe_3^+ and aromatic amino acids in the hydrophobic pocket, based on the nature of the positive charge being delocalised onto the methyl groups, and therefore are less likely to be involve in localised ionic bonding.

Identifying the active conformation of acetylcholine was important for understanding its binding. Acetylcholine is highly flexible; bond rotation along the length of its molecular framework can produce at least nine possible stable conformations and medicinal chemists strived to determine which of these shapes is exhibited in the neurotransmitter-receptor complex. This would enable design of analogues with the appropriate shape for binding.

Initially, it was presumed that acetylcholine would adopt its most stable conformation, illustrated in Figure 6.1 (a) which shows saw horse and Newman projections at the energy minimum. However, this was inaccurate: insignificant energy differences between the alternative stable conformations, such as the gauche conformation, Figure 6.1 (b) means that stabilisation energy gained from binding interactions with the receptor can more than compensate for discrepancies in conformational energies.

The illustrations in Figure 6.1 show the stereochemistry of acetylcholine: rotation about the bonds means that the amine (NMe$_3$) can be located at different position relative to the rigid C=O double bond. These conformations are associated with different energies, due to steric effects that arise as the amine NMe$_3$ and ester OMe groups occupy the same space while in the gauche conformation; as can be clearly seen when the molecule is viewed down the ester group in the Newman projections on the right-hand side of Figure 6.1.

To elucidate the active conformation of acetylcholine, rigid cyclic molecules were studied which contained the molecular framework of acetylcholine, except that the conformations in these analogues are fixed, due to the rigid cyclic component in which the bonds cannot rotate freely. If such a

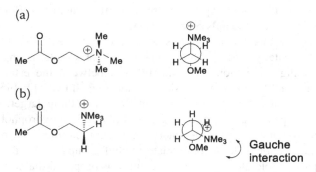

FIGURE 6.1 Molecular Structure of Acetylcholine Drawn as the Staggered Conformation (a) and as the Gauche Conformation (b); Viewing the Molecule Down the Ester Group.

molecule could bind to the receptor, it demonstrates that this particular conformation is 'allowed' for activity. By understanding the binding interactions and the nature of the pharmacophore of acetylcholine, medicinal chemists were able to develop clinically useful analogues of acetylcholine.

As previously mentioned, acetylcholine is prone to hydrolysis. This instability is an issue that had to be addressed in order to make clinically useful analogues. The primary reason for the instability of acetylcholine is due to neighbouring group participation of the nitrogen atom, making the carbonyl carbon more electrophilic. In a particular conformation, the positively charged nitrogen interacts with the carbonyl and has an electron-withdrawing effect. To compensate, the oxygen atom pulls electron density from the carbonyl carbon, making it electron deficient and therefore susceptible to weak nucleophiles, such as water.

Two strategies can be implemented to tackle the inherent instability of acetylcholine. These are the use of steric shields and electronic stabilisation. The use of a steric shield involves incorporating a bulky substituent on the ethyl bridge to protect the ester group. It does this by causing steric hindrance to the approaching nucleophiles that would otherwise initiate a hydrolysis reaction. It also interrupts the binding of esterase to inhibit enzymatic hydrolysis. The only position available for attaching a steric shield is on the ethyl bridge because, as previously mentioned, the tight fitting of acetylcholine in its receptor prohibits too much structural variation. Moreover, the inclusion of an extra methyl group to serve as a steric shield is the only option, as larger groups are not tolerated.

Electronic properties can also be used to stabilise functional groups. In the case of acetylcholine analogues, the methyl ester is converted into a carbamate to give carbachol. The lone pair of electrons on the nitrogen of the NH_2 group can be donated into the carbonyl group and therefore reduce its electrophilic character and stabilise the compound against hydrolysis. Carbachol shows good stability to chemical and enzymatic hydrolysis, but does not show selectivity between the two receptor types. Nevertheless, it was a useful treatment for glaucoma, where it could be applied locally, circumventing the issue of receptor selectivity. Glaucoma is caused by pressure building in the eye when the aqueous contents of it cannot be drained properly and can lead to blindness. Agonists cause the muscles in the eye to contract and relieve the blockage and carbachol proved a useful treatment for this condition.

The stability of the carbamate functional group and the receptor selectivity provided by the β-methyl group meant that medicinal chemists could combine steric and electronic properties to produce a stable, selective muscarinic agonist and this led to the synthesis of benthanechol, which is used therapeutically to stimulate the GIT and urinary tract after surgery.

Antagonists of the muscarinic cholinergic receptor are also therapeutically important. They work by binding to the receptor without initiating a biological response, while blocking access to the receptor by the neurotransmitter, acetylcholine. In this way, the usual biological process is suppressed. The clinical uses of muscarinic antagonists are for shutting down the GIT and urinary tract during surgery by relaxing smooth muscle. They are also useful for dilation of pupils for ophthalmic examinations and reducing glandular secretions.

The first antagonists to be discovered were natural products extracted from plants; in particular, nitrogen-containing compounds called alkaloids. For example, atropine is derived from the roots of *Atropa belladonna* (deadly nightshade) and was used in the past by Italian women to dilate their pupils, so that they would appear more beautiful, which is where the name 'belladonna' comes from. Atropine has been used clinically to reduce gastrointestinal motility. As typical for most natural products, atropine exists as one enantiomer, but racemization occurs when it is extracted into solution. This compound binds to the cholinergic receptor on account of a charged nitrogen atom when protonated, and an ester group at the required distance apart. Atropine acts as an antagonist because the molecule is larger than acetylcholine and has different binding interactions with the receptor, so does not induce the same conformational change on the receptor as acetylcholine, so the receptor is not activated.

Atropine is a tertiary amine rather than a quaternary salt and so can pass the blood-brain barrier as the free base, become protonated, and then antagonise muscarinic receptors in the CNS. This can cause hallucinogenic activity and restlessness. To reduce side effects on the CNS, structural analogues based on atropine were developed, such as quaternary salts of atropine, which are used clinically.

Antagonists of the nicotinic cholinergic receptor are also medicinally beneficial. They are present in nerve synapses at neuromuscular junctions and can be used as blocking agents. In the 16th century, when Spanish soldiers, known as conquistadors, invaded South America, the indigenous people used poisonous arrows in retaliation. The Indians used a crude dried extract from a plant called Chondrodendron tomentosum, which caused paralysis and stopped the heart. This extract is known as curare, consisting of a mixture of compounds, but later discovered that the active component was an antagonist of acetylcholine, which blocks nerve transmissions from nerve to muscle. This compound is called tubocurarine, and with controlled dose levels, is found to be medicinally useful for relaxing the abdominal muscles before surgery therefore a lower dose of general anaesthetic can be used, improving safety of operations. However, side effects on the autonomous nervous system meant that better drugs had to be developed.

The design of the drug atracurium was based on the structure of tubocurarine. This is an improved therapeutic agent, since it is free from cardiac side effects and is rapidly broken down in the blood stream, so it can be administered as an intravenous drip. The drug is rationally designed to break down via a Hoffmann elimination reaction, which proceeds at blood pH. The positive charge on the nitrogen is lost and the molecule splits in two. This is ingenious because the particular feature of the compound that gives its biological activity promotes it deactivation.

The important structural features of the drug are the spacer: a 13 atom chain which connects the tertiary centres, the blocking units, which are cyclic structures at either end of the molecule to block access of acetylcholine, and the quaternary centres which are essential for receptor binding. If one of these centres is lost through a Hoffmann elimination, the binding interaction becomes too weak and the molecules evacuates the receptor. The Hoffmann elimination is facilitated by the ester group, which serves as an electron-withdrawing group on the carbon β to the quaternary nitrogen atom therefore increases acidity of the H on the β carbon, so is easily lost in weakly alkaline conditions and the drug becomes deactivated. As a consequence, the drug acts briefly, about 30 mins, so is administered as an IV drip for the duration of surgery. Once the drip is removed, antagonism ceases almost immediately.

There are other avenues available to target the cholinergic receptor. Once acetylcholine has stimulated a neurone, it must be hydrolysed in order to prevent reactivation of the cholinergic receptor, which would increase cholinergic effects. This destruction of acetylcholine is done by an enzyme called acetylcholinesterase. Inhibitors of this enzyme, referred to as anticholinesterases will have the same biological effect as a cholinergic agonist.

To be able to design anticholinesterases, it is necessary to understand the active site of acetylcholinesterase, and the binding interactions with acetylcholine, along with the mechanism of hydrolysis. There are two features of the active site that are crucial for substrate binding. These are intermolecular forces and positioning of catalytic residues. Ionic interactions with an aspartate residue and hydrogen bonding to a tyrosine residue form during substrate binding, and an aspartate, histidine, and serine residues are involved in the mechanism of hydrolysis.

Anticholinesterase drugs inhibit the active site of acetylcholinesterase reversibly or irreversibly, depending on the interactions with the active site. The two main groups of anticholinesterases include carbamates and organophosphorus compounds. The lead compound for the carbamate inhibiters was sourced from the natural product physostigmine, which was discovered in 1864 as a product from the poisonous calabar bean from West Africa; structure determined in 1925. This compound is still used clinically to treat glaucoma. SAR studies show that the carbamate group is essential to the activity, the benzene ring is important, and the pyrollidine nitrogen is ionised at blood pH; crucial for binding to anionic residues in the active site. The carbamate group is crucial for the inhibitory properties of physostigmine. The mechanism for hydrolysis produces a stable carbonyl intermediate which is the rate-determining step. Molecular structures for some of these compounds are given in Figure 6 of the Supporting Material*. Due to serious side-effects, its medical uses are limited, so analogues have been made that retain these important features.

ADRENALINE AND THE PERIPHERAL NERVOUS SYSTEM

In addition to the cholinergic systems that rely on the neurotransmitter acetylcholine, the peripheral nervous system also includes the adrenergic system, which uses adrenaline and noradrenaline as chemical messages. The neurotransmitter noradrenaline is released by the sympathetic nerves, which supply smooth muscle and cardiac muscle with stimulus. Adrenaline is a hormone released along with noradrenaline from the adrenal medulla located in the kidneys and circulates in the blood supply in order to reach remote organs. Activation of adrenergic receptors promotes physiological effects, such as increasing heart rate and expanding lung capacity.

Tissues tend to be under a dual control system, where noradrenaline has the opposite effect to acetylcholine. Both the cholinergic system and the adrenergic system have a background level of activity and the overall response depends on which stimulus is predominant. An additional feature of the adrenergic system is the facility to release the hormone adrenaline during the 'fight or flight' response to prepare the body for immediate action by stimulating the heart and dilating the blood vessels to the muscles in times of perceived danger or stress, while shutting down digestion etc. the CNS also contains adrenergic receptors; noradrenaline has an important role in sleep, emotion, temperature regulation, and appetite.

There are two different types of adrenergic receptor, known as α- and β-adrenergic receptors, which are G-protein coupled receptors, differing in the type of protein that they are coupled to. Moreover, these two receptors have different sub-types with a different distribution around the body and have subtle structural differences that can allow for selective drug design to

---

*   Available at www.routledge.com/9780367644031

target specific organs. The neurotransmitter noradrenaline and hormone adrenaline both activate the adrenergic receptors and belong to a group of compounds called catecholamines. These compounds consist of a catechol ring (1,4-benzenediol) linked to an alkyl amine chain and are synthesised from the amino acid tyrosine. They are metabolised via enzymatic pathways.

Binding to the adrenergic receptor involves the important function groups on catecholamine. Structure-activity relationships show the importance of the alcohol groups and the phenol catechol ring, as well as the ionised amine needed for binding roles in the receptor. The alcohol group is involved in hydrogen bonding; indicated by the R-enantiomer of noradrenaline being more active than the S-enantiomer. Compounds without the secondary alcohol group, for example dopamine, have greatly reduced activity, showing that the alcohol is important, but not essential for binding. The amine is normally ionised, being protonated at physiological pH, needed for ionic bonds to the Asp-113 anion in the receptor. Activity drops with substituents on the tertiary amine/quaternary salt.

The SAR studies also demonstrate features that introduce a level of selectivity between the α and β receptors. Adrenaline has the same potency for each type of receptor, but noradrenaline has greater affinity for α receptors, which suggests that N-alkylation has implications for receptor selectivity. Indeed, increasing the size of the N-alkyl substituent results in an increase in activity for the α receptor and a decrease in potency for the β receptor. This is due to structural differences in the receptor binding sites: the α receptor contains a hydrophobic pocket into which the alkyl substituent can fit, whereas this pocket is absent in the β receptor binding site. Furthermore, extension of the N-alkylated substituent to include a polar functional group introduces favourable binding interactions, resulting in a dramatic increase in activity. For example, this can be achieved by adding a phenol group to the alkylated amine. This led to the development of adrenergic agonists, with selectivity.

The treatment of asthma demonstrates the usefulness of adrenergic agonists. Drugs specifically designed to target β_2-receptors can be used to relax smooth muscle. In the treatment of asthma, this causes dilation of the airways, and because β_2-adrenoreceptors predominate in the bronchioles, this allows for selectivity. Originally, adrenaline was used to treat asthma attacks, but because there was no selectivity for β_2-receptors, it caused stimulation of adrenergic receptors around the body and particularly had cardiovascular side-effects. Elevated heart rate increased the bodies demand for oxygen which counteracts the effects of opening the airways. The use of adrenaline is restricted to emergencies now.

The development of isoprenaline, where a bulky isopropyl substituent is added to the nitrogen, provided selectivity towards β-receptors over α-receptors, but still stimulated β_1-receptors in the heart and the problems with short duration of activity continued. Metabolic enzymes rapidly degrade these drugs to form an inactive ether. To make the drugs more resistant to metabolism, attempts were made to modify the Meta phenol group, so that the drug would not be recognised by the enzymes. Due to the importance of this group to activity, the modifications that were made needed to preserve the hydrogen bonding interactions with the receptor binding site in order to retain biological activity. A variety of different substituents were tested to understand the binding role of the phenol, for example the size of the group and electronic effects were varied and screened. Carboxylic acid groups had no activity and esters/amides were β-receptor antagonists. This research led to the discovery of salbutamol, where the hydroxymethylene group is used. The similarity between these structures is illustrated in Figure 7 of the Supporting Material*.

Salbutamol has the same potency as isoprenaline, but is 2000 times less active on the heart, and with a duration of action of 4 hrs; not being recognised by certain metabolic enzymes. Salbutamol became a market leader for treating asthma in 26 countries. It was marketed as a racemate, but the R-enantiomer is 68 times more active than the S-enantiomer, which also accumulates in the body to give side-effects. As a result, the R-enantiomer (levalbuterol) was marketed.

Salbutamol, patented in 1969, is a great example of drug design in action. The main problem with its predecessors was that metabolism was limiting duration of action. It was discovered that the OH group in the 3-position becomes methylated during metabolism and activity is lost as a consequence. This therefore may be regarded as a pharmacokinetic problem. By replacing the OH group with a CH_2OH group, binding interactions were preserved, but importantly metabolism was much slower because the terminal hydrogen of CH_2OH is less acidic, increasing the duration of action to 4 hrs. This revolutionised the treatment of the disease, which has greater than 150 million sufferers worldwide.

DRUGS OF ABUSE

The hormone adrenaline is closely related to a range of vital neurotransmitters, each of which have an important role in determining mood and behaviour. Intriguingly, all of these neurotransmitters are structurally related, with only subtle differences to optimise their binding to specific receptors. Noradrenaline is the most closely related to adrenaline, and is responsible for the amount of stimulation experienced. Dopamine is associated with the sense of 'reward' when performing a task, and serotonin is involved in mediating

---

*   Available at www.routledge.com/9780367644031

the biochemical processes of feeling good; so called 'euphoria'. In order to carry out their roles in the brain, these neurotransmitters must dock with key enzymes, which catalyse the biochemical processes associated with mood and behaviour. The structures of the neurotransmitters dopamine and serotonin are shown in Figure 8 of the Supporting Material*. Evident from structural analysis, amino-carboxylate interactions, in combination with hydrophobic interactions with the aromatic rings, are vital for the general mode of action of neurotransmitters in conjunction with enzyme binding. Other aspects of the molecular structure can be subtly varied to tune the precise activity profile. This means that this family of compounds can easily be used to control mood.

Stimulant and mind-altering drugs of this kind have been used in different forms throughout history. Mescaline, found naturally in the peyote cactus, is used by native North Americans to access a pseudo-religious trance. Another psychoactive plant extract belonging to this class of compounds is ephedrine, known since ancient China. The first fully synthetic 'recreational drug' is an analogue of ephedrine, amphetamine, first synthesised in 1887. It has been used to cheat in sports, to increase heart rate and circulation to improve the performance of the athlete, but this was risky and resulted in deaths. Interestingly, pseudoephedrine (a diastereomer of ephedrine) is used in many over-the-counter cold remedies, but has a much lower stimulant effect than ephedrine. Even though it is freely available, it is illegal in many sports, and has been implicated in many drug scandals.

Drugs have a tendency to mimic different neurotransmitters to different extents. For example, ecstasy is a strong serotonin mimic with some stimulant behaviour, but has a limited dopamine-like effect, so has a low addiction quality. Repeated use can cause overstimulation of neuroreceptors, resulting in serotonin burnout; symptoms including depression and other mood-related problems. This is because the brain becomes accustomed to large amounts of stimulus; consequently regular levels of stimulation are perceived as being in deficit to what is required to feel 'normal'.

A compound called 3,4-Methyl enedioxymethamphetamine (MDMA), widely known as ecstasy, acts as a releasing agent of neurotransmitters, namely serotonin, norepinephrine, and dopamine, which control mood and behaviour. These compounds are associated with release of the hormone oxytocin, which is implicated in the experiential qualities of the drug. Para-methoxyamphetamine (PMA), like ecstasy, is serotonergic but takes longer to achieve the same effects. It may be sold instead of, or mixed with, MDMA because it is synthesised from more readily available starting materials (anisole) rather than the natural plant extract safrole, as with MDMA. What are the consequences of this? PMA is slower acting than

---

*   Available at www.routledge.com/9780367644031

MDMA; it takes much longer than 30 min for any effect and 3 hrs to reach maximum plasma levels; MDMA takes half this time, but the half-lives of the two are similar. Consequently, multiple dosing is common when the expected effects fail to occur quickly. Furthermore, the potency of PMA is 5–10 times that of MDMA; acting to both release serotonin and inhibit breakdown of the neuro-transmitter, so low doses (plasma levels of 0.5 mg L^{-1}) cause serious elevated body temperature, resulting in overdose death. To put this in perspective, a typical 100 mg tablet of 'ecstasy' comprising 50 mg PMA will give a plasma PMA concentration of ca. 0.25 mg L^{-1} in other words two doses are likely to be fatal. Given that the user expects the rapid results of MDMA, when this fails to happen, multiple dosing will be a temptation.

The relatively simple structures of neuroactive drugs, combined with their straightforward synthesis and easy chemical manipulation, has led to the recent proliferation of psychoactive drugs which were effectively 'legal' because they have not been specifically outlawed. The most recent of these to gain wide-scale use and attention is mephedrone, which has since been made illegal. The structures of a selection of neuroactive drugs, and the molecular structure of morphine are given in Figures 9 and 10 of the Supporting Material*.

The subtlety of neurotransmitter structure leads to the possibility of a number of side effects and these are observed to greater and lesser extents for all drugs in this class. Neurotoxicity results from overstimulation of the neuroreceptors; the body becomes more adept at metabolising this type of compound to eliminate it, so eventually much higher doses are needed to achieve the desired response. Cardiotoxicity is another serious potential problem, which can result from stimulation of adrenal receptors in the heart. The extent of these problems is often unknown for these types of drugs be-cause they have never gone through any clinical trials; instead going straight from pre-clinical development to mass human consumption and the potential hazards of this may not be known for a number of years.

Natural extracts from the poppy plant have been used as effective pain killers, as well as recreationally throughout history. Opiates are not the only compounds used for pain relief, for example aspirin belongs to a different class of analgesics, and works by different mechanisms to cure different types of pain. For a long time, medicinal chemists have searched for a non-addictive analgesic based on the opiate structure, but this has proven difficult. Opium alkaloids are the nitrogen-containing natural products extracted from the opium poppy, *Papover somnfierum*. These compounds are simplistic in structure and several thousands of these alkaloids have been isolated and

---

*   Available at www.routledge.com/9780367644031

characterised to give a huge library of biologically active compounds with potential as lead compounds in many fields of medicinal chemistry.

Historical uses of opiates have been recorded as early as 2000 years ago in ancient China, and because of its properties, the ancient Greeks dedicated the opium poppy to Thanatos (god of death), Hypnos (god of sleep), and Morpheus (god of dreams). Later, physicians prescribed it for a range of afflictions, such as headache, epilepsy, vertigo, and asthma, as well as use as a sedative. Use as an analgesic came much later. In the 16th century, Paracelsus introduced a preparation of opium called laudanum. The drug became increasingly popular, due to its wonderful effects. However, its addictive qualities led to problems which arose with long-term use; observations by doctors quoting 'great intolerable distress, anxieties, and depression of the spirit'. Of course, these were symptoms of withdrawal.

Opiates were first marketed in Britain by Thomas Dover. Dover had a colourful past, with an early career in piracy and famed for rescuing the marooned sailor Alexander Selkirk from an uninhabited island, which is thought to be the inspiration for Dafoe's *Robinson Crusoe*. Dover later in life took up medicine and one of his products was a powder containing opium, liquorice, saltpetre, and ipecacuanha, which is an emetic to make consumers sick if they took too much.

Godfrey's cordial from the 18th century was another popular remedy, containing opium, treacle, and sassafras, used for treating rheumatic pains. This preparation was freely available in most grocery shops without prescription or restriction, despite many people becoming addicted. As late as the 20th century, opium was viewed as a legitimate substance, like tobacco and tea, but by 1920, the use of opiates was prohibited, and could only be prescribed by doctors if patients demonstrated a legitimate medical need.

Obsolescence of opium ensued after the raw material was superseded when purified, semi-synthetic, and synthetic opiate analogues were developed. Morphine was the first pharmaceutical to be isolated from a plant extract. It is the main alkaloid present in opium and is responsible for the analgesic activity. It was discovered in 1804 by Friedrich Wilhelm and Adam Sertürner, who after 13 years of research, published the isolation of morphine in 1817. The advantage of the purified substance was that patients could be treated with a known dose, and avoid the dangerous effects that were associated with consuming large quantities of opium. The sale of morphine commenced in 1827 by Heinrich Emanuel Merck of Darmstadt, who was able to expand his family pharmacy into the Merck KGaA pharmaceutical company. Heroin was the first semi-synthetic opioid, and was marketed in 1898 by Bayer as a safe alternative to morphine because its lethal dose is hundreds of times above its effective dose. Morphine effects the central nervous system by binding to receptors in the brain and spinal cord, as well as to receptors present in the stomach and intestine; potentially causing harmful effects such as respiratory difficulties, coma, or cardiac or respiratory collapse.

Targeting the nervous system can be a chancy business; medicinal chemist must carefully design drugs so that side effects are minimal, but create compounds with enough potency to be therapeutically useful. Due to the subtleties of the cholinergic and adrenergic receptors, greater attention is needed in understanding structure-activity relationships because active compounds will all have similar molecular structures, and small alterations to the structure may have dramatic consequences regarding activity. The reward for creating these drugs is substantial because disorders of the nervous system can be very problematic for patients. Unfortunately, the effects of psychoactive drugs can be desirable recreationally and this can lead to the abuse of certain compounds.

Drug Design, Synthesis, and Development

7

DRUGS IN THE HUMAN BODY

When designing new drugs, medicinal chemists must take into consideration both the interactions of the drug molecule with its target and the compounds behaviour in the body. Pharmacodynamics examines how to optimise a drug binding to its target. Pharmacokinetics considers how a drug travels through the body to reach the target. A compound that has the best binding interaction is not necessarily the most medicinally useful; as efficacy depends on the drug reaching the target at therapeutic concentrations. Often, a compromise between pharmacodynamics and pharmacokinetics has to be made. Considerations for the pharmacodynamics properties of drugs has been discussed in previous chapters; optimising drug binding with the target is an essential feature of drug design. Designing molecular features to facilitate a drugs journey to the target is equally important. There are five main aspects of pharmacokinetics to take into consideration when designing drugs: absorption, distribution, metabolism, excretion, and toxicity; abbreviated as ADMET.

Drug absorption refers to the route or mechanism by which a drug reaches the blood supply, which is also dependent on how the drug is administered. The preferred method of administration is the oral route because it is the least intrusive, hence is used most commonly. When taken orally, the drug is delivered to the gastrointestinal tract (GIT); first entering the stomach, where it is subjected to gastric juices and hydrochloric acid. These chemicals, used to digest food, will also degrade drug molecules, so for a drug to be effective orally, it must be tolerant to these conditions.

In order to enter the blood stream, the drug must pass the cells lining the intestine. Once the drug has traversed the cell membranes, it can readily enter capillaries through pores that exist between the cells of the blood vessels. Upon entering the capillaries, the drug is on route to the liver, which contains enzymes designed to intercept and modify foreign chemicals in order to make them easier to excrete. This process is known as drug metabolism. This introduces the necessity for an orally active drug to be metabolically stable as well as resistant to the enzymes and hydrochloric acid of the GIT, and this presents a challenge to medicinal chemists designing new drugs.

In addition to metabolic stability, a drug must have the correct balance of water solubility versus fat solubility. Hydrophilic drug molecules that are too polar may fail to pass through the fatty cell membranes of the gut wall. A drug that is non-polar could be too hydrophobic and poorly soluble and dissolute into globules in the gut. In both these instances, absorption would be poor. To help balance the dual requirements of water and lipid solubility, the amine functional group is often incorporated into drug molecules, and amines can often be involved in target binding also. Amines act as weak bases, so are partially ionised at blood pH and readily equilibrate between their ionised and non-ionised forms, which means that they have good water solubility, but will also pass cell membranes as the non-ionised form.

Typically, the hydrophilic/hydrophobic character of a drug determines how readily it will be absorbed. The size of the molecule also has an influence on drug absorption. As well as physical constraints, larger molecules are likely to contain more polar functional groups, so in principle are less-readily absorbed through cell membranes. Certain criteria have been postulated for an orally active drug to be effective, known as Lipinski's rule of five: so-called because the numbers involve multiples of five. The drug's molecular weight must be less than 500 g mol^{-1}, it must have no more than five hydrogen bond donor groups, and no more than ten hydrogen bond acceptor groups, and have a log P value less than five (measure of hydrophobicity).

Polar groups that break these conditions tend to be poorly absorbed and need to be administered by injection. Alternatively, drugs can be designed to hijack carrier proteins. If the drug bears a structural resemblance to some of the polar biomolecules used in the cells natural biosynthetic pathways, the appropriate carrier protein embedded in the cell membrane can be utilised to transport the drug molecule across the cell membrane and into the cell. For example, fluorouracil is transported by carrier proteins for the nucleic acid bases thymine and uracil. Also, small polar drug molecules with molecular weights less than 200 g mol^{-1} can pass through small pores between the cells lining the gut wall, or they can be carried in vesicles (pinocytosis).

Drug distribution is the next step in the pharmacokinetics of a drug. Once the drug has diffused into the capillaries, it is speedily distributed around the blood stream, and then more slowly distributed to various tissues and organs. Drug molecules do not have to cross cell membranes in order to leave the blood system and enter tissues. Capillaries contain small pores between cells through which drug-sized molecules can pass, but not important plasma proteins. If the drug target is an extracellular receptor, the drug can find the target, bind to it, and perform its function. Drugs that have targets within the cell, such as enzyme inhibiters, or drugs acting on nucleic acids, must leave the extracellular aqueous fluid and traverse cell membranes to access their targets. To do this, the drug must be sufficiently hydrophobic, or be able to utilise carrier proteins, or be taken in by pinocytosis. If a drug is excessively hydrophobic, it can become absorbed into fatty tissue, thus lowering the concentration available to bind to the target. Likewise, drugs that are ionised may be less available in the blood supply if they become bound to macromolecules; for example they can become irreversibly bound to blood plasma proteins.

Designing a drug that can pass the blood-brain barrier is a challenge. Capillaries that supply the brain are lined with tightly fitting cells that do not contain pores. Furthermore, they are coated with a fatty layer formed from neighbouring cells, which presents an additional barrier to the passage of drugs. Consequently, polar drugs, example penicillin, cannot enter the brain. However, this has the advantage for designing drugs to be more polar to avoid CNS side effects. The ability to cross the blood-brain barrier is important for the activity of opiates. Polar groups must be temporarily masked using the pro-drug approach, or design compounds that utilise pinocytosis in order for the compounds to reach the site of action.

The extent of drug metabolism is another contributing factor to the pharmacokinetic properties of a drug. Drugs may be reactive with a range of metabolic enzymes, which aim to modify the foreign molecule into a structure, known as the metabolite, which is more readily excreted. These transformations may lead to loss of activity of the parent drug, or the metabolite may even have a different activity and lead to side effects or toxicity. Hence, medicinal chemists must design drugs consciously not to produce unacceptable metabolites. Indeed, it is a requirement that all metabolites of a drug are characterised before it is approved, including stereochemistry and tested biologically. On the other hand, metabolic pathways can be important for prodrug approaches.

The body has systems in place to remove foreign chemicals. Polar molecules can be quickly extracted by the kidneys, but non-polar molecules must undergo metabolic transformations to make them more polar before they can be excreted. Non-specific enzymes, such as cytochrome p450 enzyme in the liver, can add polar functional groups to a wide range of compounds to improve their water solubility and make them more easily excreted through the kidneys in urine. A different series of enzymatic reactions might reveal polar functional groups that are masked. For example, there are enzymes that can demethylate a methyl ester to unveil a more polar hydroxyl group. This metabolite can be excreted more efficiently.

These reactions are classed as phase 1 reactions and generally involve oxidation, reduction, and hydrolysis; typically taking place in the liver. These reactions also tend to involve characteristic functional groups. Some of the structures most prone to oxidation are N-methyl groups, aromatic rings, and the terminal positions of alkyl chains. Reduction with reductase enzymes will occur with nitro, azo, and carbonyl groups, while amides and esters are prone to hydrolysis by esterase enzymes. Knowledge of the metabolic reactions that are possible for different functional groups enable medicinal chemists to predict the metabolite that a drug is likely to produce. A drug may undergo several of these reactions and yield may different metabolites. Understanding this chemistry helps with drug metabolism studies to identify exactly which metabolites are formed and hence explicate the future prospects of the drug in regards to safety.

Stereochemistry is an important consideration when designing drugs, as chirality can have important implications for drug metabolism. Metabolic enzymes can often distinguish between two enantiomers, such that each can undergo different reactions. This means that testing each enantiomer separately during screening is necessary and may mean that the design of a drug may need an asymmetric synthesis.

Within the liver, another series of reactions, phase 2 reactions, proceed and regularly involve conjugation reactions, where a polar molecule is attached to the drug at a polar functionality, which may have been placed there by a previous phase 1 reaction. The resulting conjugate will be more polar, thus more readily excreted in the urine or bile.

Metabolic stability is an important property of a drug, given that metabolism usually results in loss of activity, or even result in toxicity. This also has implications for dosing. Another issue is that the activity of metabolic enzymes varies between individuals, particularly for cytochrome p450 enzymes. The difference in the activity profile of these enzymes between patients will affect how fast a drug is metabolised and therefore the dose of the drug must be monitored, depending on the rate of metabolism. Indeed, different countries can have different recommended dosing levels of a drug because of the differences in metabolism between populations. Pharmacogenomics concerns with screening different populations; how genetic variation affects individual's response to a drug, and therefore gauge the appropriate dosing level.

Metabolism by Cytochrome p450 enzymes is also affected by other chemicals. For example, other drugs and even certain foods can influence these processes. For certain drugs, brussel sprouts and cigarette smoke can enhance activity, whereas grapefruit juice suppresses it. Because a person's diet can influence how a drug is metabolised, recommendations are usually given regarding what foodstuff should be eaten while taking the medication.

For example, the immunosuppressant drug cyclosporine has improved activity if taken with grape fruit juice; being less speedily metabolised, but if the antihistamine terfenadine, which is a pro-drug, is taken with grapefruit juice, inhibition of metabolism that produces the active compound means that terfenadine persists in the body and can have cardiovascular side effects; hence, the active ingredient is now administered directly and is marketed as Allegra.

Drug-drug interactions, where one drug affects the activity of another, can arise when certain medicines are taken together. A common scenario is when a person takes antibiotics while on other medication. Many antibiotics inhibit cytochrome p450 and this can have consequences for any other drugs being taken. As a result, new drugs are tested to examine whether cytochrome p450 is inhibited or activated because of these challenges.

Drugs need to be designed that live long enough in the body to perform their function, but are metabolised quick enough so as not to cause side effects. When taken orally, drugs must pass directly to the liver where a percentage will be metabolised before the drug continues to circulate the blood stream. This is known as the first-pass effect. This can affect whether or not a drug reaches the target at therapeutic concentration, hence, sometimes drugs must be administered intravenously.

Drug excretion from the body can occur in several different ways. Drugs can be diverted from the liver to the intestines in bile and excreted in faeces. Volatile drugs can be exhaled from the lungs, and up to 10–15% of a drug can be lost through the skin in sweat. Predominantly, drugs are metabolised and excreted via the kidneys. The function of the kidneys is to filter the blood of waste chemicals, which are subsequently removed in the urine. Blood enters the kidneys via the renal artery, which divides into many capillaries, each of which forms a knotted structure called a glomerulus that fits into the opening of a duct called a nephron. Blood entering the glomeruli is under pressure, causing plasma to be forced out through the pores in the capillary walls and into the nephron, along with the dissolved drug and metabolites contained within. Components that are too big to pass through the pores, such as plasma proteins and red blood cells, remain in the capillaries within the remaining plasma. This is a filtration process, so whether the drug is polar or hydrophobic does not affect the efficiency of passage into the nephron. However, not all drugs and metabolites will be excreted with equal efficiency, depending on the hydrophobicity of the compound. Some of the compounds that have passed into the nephron will be reabsorbed into the rich network of blood vessels surrounding the nephron before reaching the bladder. As water from the filtered plasma is reabsorbed, drugs and other compounds become concentrated in the nephron. This produces a concentration gradient and hydrophobic compounds may diffuse across the cell membranes of the blood vessels and become reabsorbed in the blood. Polar compounds are retained and dissolve in the urine to be excreted.

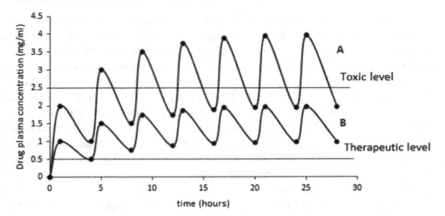

FIGURE 7.1 Two Drug Dosing Regimens Showing the Change in Drug Concentration in Plasma to Illustrate the Therapeutic Dosing Window.

The pharmacokinetics of a drug, which can be broken down categorically as ADMET, are the major determining factors that govern drug dosing. There are many pharmacokinetic variables involved in drug distribution; many dependent on the route of administration, so estimating the correct dosing regimen can be a challenge, where considerations for the concentration of the drug in each dose and the frequency of administration should be taken into account. Drugs are taken at regular time intervals to maintain a fairly constant level in the blood: not too low as to be ineffective and not so high to be toxic. This equilibrium is known as the therapeutic dosing level.

Figure 7.1 shows how the therapeutic dosing level is established. Dosing regimen A quickly reaches the therapeutic level, but continues to rise at a steady state to reach a toxic level. Regime B involves half the amount of drug at the same dosing frequency and takes longer to reach the therapeutic level, but the steady state remains below the toxic level, within the therapeutic window.

Complications in drug dosing arise due to differences in age, sex, and race etc. between different patients. For example, people who are obese present a particular challenge because it is hard to predict how much of the drug will be stored in fat tissue. The precise time of drug dosing is another aspect that is difficult to predict because metabolic reactions proceed at different rates throughout the day; medicines to be taken before bed, when metabolism is slower during sleep, will probably have a slower activity than medicines that need to be taken multiple times throughout the day. One crucial variable for determining how frequently drug dosing is needed is the drugs half-life ($t_{1/2}$), which is the time taken for the blood concentration of the drug to fall by half.

This depends on excretion and drug metabolism processes, which does not occur linearly with time. Half-lives may vary from a timescale of just a few minutes, such as some of the opiate analgesics, to a number of days, as with diazepam (Valium), where it can take over a week to recover from the effects. In order to maintain a steady state concentration in the blood, the dosing must be complementary with the rate of excretion and metabolism, thus, it is important to know the half-life of the drug. Generally, the time taken for a drug to reach its steady state concentration is about six times its half-life. For example, in Figure 7.1, the drug has a half-life of 4 hrs and is dosed at 4-hr intervals. Notice that half the level is at a maximum after each dose and falls to a minimum before the next dose is provided. It is important to ensure that the blood concentration does not fall below the therapeutic level, or exceed the toxic level, in other words, remains in the therapeutic window. The plasma level of the drug at steady state depends on the size of the dose given as well as the frequency; therefore, during clinical trials, blood samples are regularly taken from patients in order to establish safe dosing levels.

Drug tolerance is another aspect of consideration. With some drugs, the effects may diminish with repeat dosing, therefore dosing needs to be increased. This may arise with an increase in production of metabolic enzymes in response to the drug, or the target may adapt to the presence of the drug, for example antagonism of a receptor may lead to biosynthesis of more cellular receptors, hence more drug would be needed. These effects can lead to chemical dependency, where the patient requires the drug in order to feel normal; otherwise withdrawal symptoms ensue. This relates to drug tolerance in that, if more cellular receptors are produced, the cells become oversensitive to the substrate; be it neurotransmitter or hormone and it can take weeks for cellular mechanisms to breakdown the excess receptors, during which time the patient will experience discomfort from withdrawal symptoms. Tolerance to drugs can also lead to increased risk of side effects and overdose and the danger of the user incurring harm is far greater.

HOW DRUGS ARE MADE

Medicinal chemists strive to make compounds that serve as remedies for poor health. In order to achieve this, a thorough understanding of the cellular and molecular biology of the ailment is required as well as an extensive knowledge of synthetic organic chemistry. First, a viable target must be identified for the disease; next identification of a hit compound, which can be achieved by molecular modelling of hit-target binding interactions, and screening a range of related compounds in the laboratory, or computationally to find the compound with the optimum structure-activity relationship. Lead optimisation is the next phase of drug development, where the binding properties of the hit are further optimised through modification of the molecular structure to maximise drug-target interactions, but also taking into account the pharmacokinetic properties of the drug as well.

Organic synthesis is of paramount importance for creating new drugs. Knowledge of the chemistry of different functional groups enables medicinal chemists to select particular chemical reactions to use to combine molecular fragments and build a predetermined molecular structure of a drug molecule, based on the outcome of computational analysis and lead optimisation processes. Similarly, the required functional groups that are needed for binding to the target can be introduced to the molecular structure by selecting the appropriate chemical reactions for adding a new functional group, or transforming an existing functional group into the one that is required. The chemical reactions needed to produce the novel compound are set out chronologically to yield a recipe for making the new drug, and this is referred to as a synthetic strategy.

The molecular framework of drug compounds commonly comprise aromatic, or heteroaromatic ring structures. These can serve as a scaffold for holding the functional groups involved in binding interactions in the correct positions, and can be involved in binding as well. Aromatic and heteroaromatic compounds have a diverse chemistry and offer many opportunities for chemical reactions to introduce new functional groups, or amalgamate molecular fragments, needed to build a target compound.

Benzene commonly reacts with electrophiles via electrophilic substitution reactions. For example, nitration, sulfonation, and halogenation reactions, which all require a Lewis acid catalyst; an acceptor of electrons, are useful ways of introducing new functional groups to aromatic rings. Another useful example for synthesis would be the Friedel-Craft reactions (acylation and alkylation) because these form new carbon-carbon bonds, useful for combining molecular fragments.

Reactions of substituted benzenes depends on the functional groups already present on the ring. Regioselectivity, which governs where on the ring the reaction takes place, depends on the electronic effects of the substituents and their position on the ring. Activating groups are typically +I and +M groups, where positive induction and positive mesomeric effects donate electrons into the aromatic ring system. Deactivating groups withdraw electrons from the ring (−I and −M) and destabilise the ring, making it less reactive. Activating groups direct the reaction to the ortho and para position (depending on steric factors of the substituted benzene), while deactivating groups direct to the meta position, except for halogens, which are o/p directors. During synthesis, it is therefore important to add the substituents in the correct order to achieve the desired regioslectivity.

Heteroaromatic compounds are more reactive than benzene and will readily react with electrophiles at the C2/C5 position, due to more resonance structures being available to stabilise the positive charge of the transition state. Six-membered heteroaromatics, such as pyridine, will also undergo nucleophilic substitution at C2 and C4 because the nitrogen acquires a negative charge through resonance. Methyl pyridines do not behave in the same way and instead react at the methyl group due to the negative inductive (-I) effect from nitrogen. These compounds therefore offer interesting alternative ways of assembling a target compound.

When it comes to building a drug molecule, carbonyl chemistry offers a rich variety of reactions that can be used to create new functional groups, or make new carbon-carbon bonds, which are so important for combining different molecular fragments together that comprise the final target compound. Ketones and aldehydes are versatile functional groups and will undergo a diversity of reactions. The electronegative oxygen polarises the C=O bond, making the C susceptible to nucleophiles and the αH easily deprotonated with a base. Predominantly, ketones and aldehydes will undergo nucleophilic addition reactions; this includes reduction with hydrides to an alcohol and addition of nucleophiles to the carbonyl carbon. A range of nucleophiles can be used, including alcohols, amines, and cyanide, which are particularly useful in synthesis to introduce new functional groups, and Grignard's reagent to make a new C–C bond. All of these basic organic reactions of carbonyls are invaluable for building a target compound.

The α hydrogens of a ketone can readily be deprotonated with a base: α substitution reactions involve replacement of the αH with another group via an enolate. Useful reactions in synthesis include aldol condensations, where two ketones are combined in an alkylation reaction. In each case, a new C2–C3 bond is formed. This is a common strategy in organic synthesis because of the versatility of the C2–C3 bond in making new compounds. Ketones can also be used as a starting material for making alkenes, which provides an opportunity for a whole new set of organic reactions and alternative synthetic routes towards producing a target molecule.

Often it may not be feasible to introduce the required functional groups straight away. This could be due to the types of reactions being implemented in the synthesis and reaction selectivity for different functional groups. As a result, reactions known as functional group interconversions may be required towards the end of a synthetic strategy to unveil required functional groups. Reduction of organic functional groups is a common reaction used for this purpose.

In many cases, the stereochemistry of the drug molecule is of critical importance. Consequently, the synthetic strategy needs to account for stereochemistry in regards to the starting materials being used and the stereoselectivity of the chemical reactions utilised in the synthetic strategy. Often an asymmetric synthesis is required, which will involve specific reagents and chemical processes.

The synthetic strategy for making a drug can be devised by working backwards from a given target molecule. This approach is called retrosynthetic analysis. In this method, organic syntheses can be planned by breaking down a target molecule into smaller building blocks that are commercially available, known as readily available starting materials. This is done by the imaginary breaking of bonds, referred to as disconnections, indicated by an open-ended arrow to produce synthons, or by functional group interconversions. Synthetic equivalents are then used to accomplish the forward reaction in a synthesis.

An example RSA is given in Figure 7.2. Bayer started to explore azo dyes as antibiotics and this lead to the development of sulfa drugs, such as prontosil red (2.). Despite this drug being effective *in vivo*, it did not affect bacterial cultures *in vitro*, suggesting that an active pharmacophore is produced in the body; later discovered to be structure 1. Consequently, further synthesis from 1. was no longer necessary. These compounds can be easily made and a range of analogues were synthesised to find the most effective, and at the same time reveal the structure-activity relationships of this class of drugs.

The synthesis starts from benzamide, where the amide group acts as an o/p director and directs the reaction with the sulfa group on to the 4-position of the ring. Chlorine is then substituted by ammonia, followed by base-catalysed cleavage of the carbonyl group to produce the active compound. A functional group interconversion is implemented, transforming the amine to the azo group. Elecrtrophilic aromatic substitution reaction, supported by the electron-donating properties of benzamine, produces the azo dye.

Once the synthesis of a drug is complete, the compound is not yet ready for consumption. Thorough purification steps are required to clean the organic compound. Chromatography is a commonly adopted technique for separating a target compound from the rest of the matrix. In chromatography, the column is packed with a solid stationary phase consisting of very fine particles and a solvent is selected, referred to as the mobile phase, in which the target compound dissolves. The desired target product will have a different affinity for the stationary phase than any by-products or unreacted material, and therefore will have different retention times on the column. The product elutes from the column at a different time to the contaminants, so can be isolated. The greater the length of the column, the greater the degree of separation, as in the case of high-performance liquid chromatography (HPLC) and gas chromatography (GC), used for volatile compounds, and even separation of stereoisomers is possible.

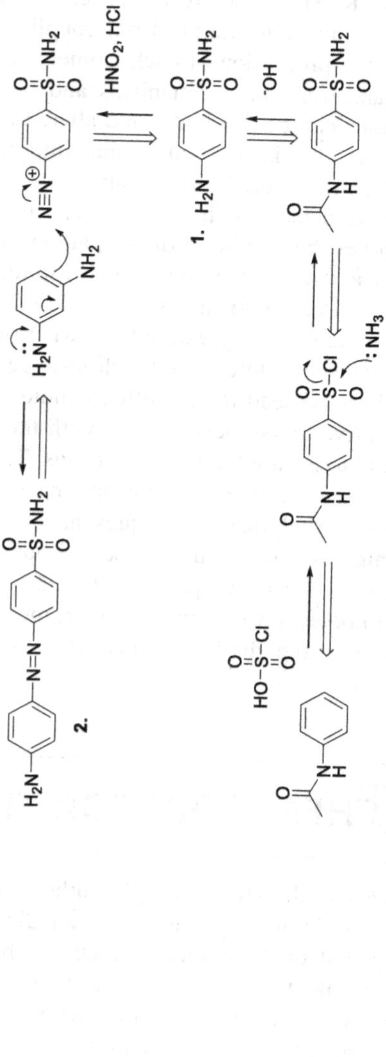

FIGURE 7.2   RSA for the Antimicrobial Drug Prontosil Red.

Chromatography can be coupled to mass spectrometry, MS, which enables the relative molecular mass of the analytes to be accurately determined and help to characterise product and contaminants. However, this will not give information on isomers. Spectroscopic techniques are required for complete characterisation. Infra-red spectroscopy is a quick and useful technique for identifying functional groups that are present in the molecule, but does not yield complete structural information and this is where nuclear magnetic resonance (NMR) spectroscopy is needed. This enables accurate structural characterisation, and information from coupling constants and integration can determine the proportion of each isomer present.

Understanding organic reaction mechanisms and the step-by-step processes by which molecules can be combined and altered enables chemists to design novel compounds. This is of fundamental importance to drug discovery because new drug molecules can be rationally designed to interact with a biological target associated with a disease and therefore have a biological effect that improves the condition of the patient. Subtle modification of molecular structures and lead compound optimisation can be done by implementing techniques in organic synthesis to yield the best possible drug properties of a lead compound. A key example of where this is important is asymmetric synthesis, where a stereospecific drug-target interaction is required, which could otherwise lead to side effects; note the case of thalidomide. While it is imperative to synthesise drugs with the least possible side effects, it is also essential to ensure the purity of drugs. This is the role of the analytical chemist, who has the important occupation of screening for and eliminating contaminants. Analytical techniques have improved greatly in modern times; contaminants can be traced below nano-gram levels, and methods such as spectroscopic techniques can be used to characterise the contaminants. As technologies advance in the future, medicinal chemists will be well equipped to manage the challenges that are presented to the field of medicine in the coming decades.

MEDICINAL CHEMISTRY FOR THE FUTURE

Ever since the discovery of the atom through Rutherford's alpha-scattering experiments in 1911, our knowledge and understanding of chemistry has catapulted the development of the subject, which has been instrumental in facilitating many advancements in our modern society. The nature of atoms and understanding how they interact and bond with one another is fundamental to chemistry. Having an appreciation of the three-dimensional properties of molecules is essential to structure and bonding; molecular geometry and isomerism are important features of drug molecules and govern how drug molecules bind to their target through intermolecular interactions.

To design and synthesise drugs, the interactions of molecules in terms of organic reactions and their relevance to synthesis is foundational to medicinal chemistry. Computational modelling has the capability to reveal potential molecular structures which might interact with a biological receptor in a way that is medicinally beneficial. The challenge for organic chemists is then to use their knowledge of the subject to design a strategy to synthesise the target molecule (TM) from readily available staring materials (RASM), which ideally are compounds that are inexpensive and commercially manufactured. In general, many similar structures are synthesised in what is referred to as a compound-library. Structures produced as potential hits are screened and those compounds found to have good structure-activity relationships go on for lead optimisation.

In order to design a target molecule for a biological receptor, medicinal chemists need to understand the molecular basis of diseases to be able to identify targets. Targets are often proteins; hence, an understanding of protein structure is important. Knowledge of the amino acid sequence and the three-dimensional tertiary or quaternary structures is vital for building a model to investigate possible binding interactions of ligands, which could potentially lead to drug development. Their importance in catalysis, cellular recognition, and structure means that proteins have a high likelihood of being possible targets for a disease. It is also important to understand genetics because genes contain the underlining information for protein synthesis, and often the mechanisms of a disease, therfore present other potential targets, such as gene therapy. Indeed, the future of medicinal chemistry is likely to be based more around genetics as our understanding of the subject grows.

Many of the first cases where drugs were rationally designed against identified targets was in the treatment of viral diseases. Targets were revealed from molecular modelling, based on information from x-ray crystallography of virus components, then molecular structures could be designed that would be expected to interact with the virus in a certain way. This is a good example of where understanding the molecular basis of disease is utilised in medicinal chemistry, for example the influenza life cycle, which was discovered from structural analysis and viral genetic sequencing. Viral genetic sequencing yields the potential for synthesis of pure proteins that can be used for drug-target binding assessments.

Drugs can be designed to operate on targets in different ways, for example inhibition of an enzyme by a compound that mimics the natural substrate to block the active site, such as in the case of Ralenza™ and Tamiflu™ blocking the neuraminidase active site; an enzyme important in the life cycle of the influenza virus. Compounds can be designed to interact with a target with subtle properties of the molecule deliberated during optimisation to have the effect of agonism or antagonism, depending on what property is required. By understanding the protein three-dimensional structure-function relationship and building the drug molecule appropriately, accounting for hydrogen bonding and optimising all non-covalent interactions, a drug can be designed to either enhance or suppress, respectively, the activity that results from the natural substrate. The subtle results for correct optimisation were exemplified by Relenza™ and Tamiflu™ in terms of the method of administration. Although both structures achieved strong binding that was needed for inhibition of neuraminidase, Tamiflu was designed with more hydrophobic moieties which meant that it could be absorbed across the gut and was orally active, which is preferable to Relenza which had to be inhaled, which could lead to irritation.

Rational design of drugs to an identified target, can often take the approach of combinatorial chemistry, known as the molecular fragments technique, where sections of a compound that are discovered to bind well to a section of the target are spatially positioned and a scaffold section of the molecule is designed to hold the fragments in these positions. This is how Glaxo Smith Kline and Roche came to develop similar structures independently. The next stage is lead optimisation, where drug properties are optimised. For example, the improved oral bioavailability of Tamiflu compared to Relenza. Idealised drug criteria for an effective medicine involve: high affinity for the target, safe/well tolerated by the patient, synergistic with other drugs when used in combination therapy, and can be taken orally with minimal dosing frequency. The closer a drug adheres to these requirements, the greater its potential as a medicine.

Interestingly, Tamiflu is prepared through a complicated 15-step synthesis, involving shikimic acid, extracted from star anise as a starting material. Despite the lengthy synthesis, Roche uses 90% of the global crop of star anise to make Tamiflu because of the mass market and profitability from government stockpiling of the drug. Unfortunately, though, some influenza strains now show resistance to Tamiflu. The mechanism of mutation involves a single amino acid mutation in the neuraminidase active site, which prevents effective drug binding without loss of function of the enzyme.

The occurrence of resistance and its challenges to medicine are highly relevant to antibiotics and the treatment of bacterial infections. Indeed, resistance to antibiotics probably has greater media attention than for any other class of drugs and this is because it is such an important issue. To demonstrate how crucial antibiotics have been to developing society in the past, consider the morbidity in history prior to antibiotics. The advent of antibiotics was a cultural and medical revolution, which resulted in increased success rates of surgery and a dramatic increase in life-expectancy.

As the field of medicine advances into the future, it is vital to recognise the challenges of today; problems in developing countries with antibiotic availability, cholera, epidemics, and the risk from globalisation regarding the transmission pathogens. Antibiotic resistance in the developed world, notable examples such as multi-drug resistant tuberculosis (TB) and MRSA, are a public health crisis. The spread of harmfull viruses with the increase in globalisation is another huge challenge for medicine. As we move into the future, medicinal chemistry will need innovation; designing new drugs to battle infection. This issue has been dramatically exemplified by the coronavirus pandemic of 2020.

Paramount for designing new drugs, an understanding of the molecular basis of the disease, as emphasised so many times throughout this text, is essential for understanding the resistance mechanisms that bacteria employ. Bacteria can produce more of an enzyme that metabolises the drug, change the target enzyme active site subtly, so that the drug cannot bind effectively and therefore does not work, and decrease cell wall permeability to the drug or express more efflux pumps to remove the drug from the cell; all of these mechanisms derive from mutation. An example case study is penicillin re-sistance, where mutation led to the production of an enzyme, β-lactamase, which broke down the drug. Consequently, strategies needed to be employed to combat this problem, such as the use of clavulanic acid administered as Augmentin, to improve the efficacy of Amoxicillin. Chemists need to identify targets to address these problems. Strategies adopted in the past such as steric shields to prevent access to the β-lactamase active site, example methicillin, relied on understanding the binding interaction with the target. However, steric shields must not compromise enzyme-substrate complementarity with the target. Another approach was to search for other targets and develop drugs with different mechanisms of action, for example vancomycin acted on the bacterial cell wall via a different mechanism to penicillin. Finding alternative targets besides the bacterial cell wall was another strategy, such as in the case of tetracycline, which acts to disrupt bacterial protein synthesis.

Resistance to vancomycin is a big problem because this drug is often regarded as the last line of defence against pathogenic bacteria. Occurrence of "superbugs" as pathogenic bacteria become tolerant to a number of anti-biotics, resulting in multi-drug resistant genes on plasmids that can be rapidly transferred between bacteria and to daughter cells is a huge challenge to medicinal chemistry. Resistance arising from overuse of antibiotics, people not completing a course of medication, and trace amounts of antibiotics in hospitals are all influencing factors resulting in the creation of superbugs that develop tolerance to many kind of antibiotic. These are all habits that need to be addressed to help prolong the life-expectancy of antibiotics. In the future, there could be the possibility of identifying new targets from genome sequencing and understanding the genetic-level mechanisms of disease. This could allow for the development of a whole new class of antibiotics that pathogens are not familiar with.

Keeping one step ahead of mutation will be key to combating pathogens in the future, for example utilising a multi-drug approach to avoid selection of resistant pathogens like the case of the Influenza virus varying the NA binding site, or in the battle against MRSA. It is likely that the arms race between pathogens and drugs will never cease. The unnecessary use of drugs should undoubtedly be avoided to try and preserve the arsenal we have against pathogens for as long as possible. However, from the individual patient's perspective, being denied medicine when they feel unwell and the need to justify their access to drugs can be a challenge, and restricting drug use can also impact the profitability of pharma industry, which is critical for research and development. These ethical and philosophical considerations could affect the evolution of medicine in the future.

One of the great medical challenges to modern society is cancer. With a growing and aging population, the occurrence of the disease has skyrocketed, with one in three of us developing the disease according to the American cancer society. Cancer can be a consequence of external influences such as exposure to chemicals that are damaging to DNA, carcinogens, and viruses which carry oncogenes. These lead to mutations of DNA and if genes controlling the cell cycle are damaged, this results in uncontrolled replication of cells into tumours. This can be an age-related breakdown of control mechanisms as well as longer exposure to carcinogens. Cancer can also spread, which is known as metastasis. This is another challenge of the disease, making it harder to control.

A molecular understanding of DNA replication helps to understand the disease and assists to find ways to combat cancer. One possibility might be to look at mitosis to find targets. Understanding the molecular basis of complex diseases to identify targets and develop therapies, for example, from understanding mechanisms of the cell cycle and DNA replication, might elucidate ways in which the disease might be treated, but the high complexity means that this is an immense task. As cancer treatment advances into the future, it will be vital to find mechanism-specific ways to combat the disease because the current approaches to cancer therapy have so many problems. Traditional ways to target cancer have harsh side effects; targeting DNA directly in the case of alkylating agents and intercalaters, which act by distorting the α-helix and cross link strands to inhibit DNA replication also effects normal cells. Mustard alkylating agents and cis-platin cross-linking agent both have highly adverse side effects because of their influence on normal cells as well. Use of metal complexes has been explored to produce less-toxic analogues of cis-platin, but these still cause harm to normal cells. Hormone therapy, targeting the oestrogen receptor, is another method to fight cancer, but the lack of discrimination for normal cells still remains.

Frontiers in cancer treatment look to create therapies without the severe side effects of traditional chemotherapy. The methods focus more on target specificity, such as the use of antibodies in antibody directed enzyme pro-drug therapy (ADEPT). This technique harnesses a "seek and destroy" approach where the specificity of antibodies is utilised to take the drug directly to the site of action, thereby minimising the effect on normal cells. Ingenious strategies such as this will be the focus of cancer therapy in the future.

Neurodegenerative diseases, which are associated with old age will become increasingly common in the future with aging populations. Conditions such as Alzheimer's disease result from the breakdown of normal biochemical processes in the neurones of the brain. Beta-site amyloid precursor protein cleaving enzyme (BACE), also known as beta secretase is an aspartic acid protease that is important in the formation of myelin sheaths in peripheral nerve cells. Elevated levels of this enzyme are present in patients with late-onset sporadic Alzheimer's disease. Generation of amyloid-β peptides, which aggregate in the brain of Alzheimer's patients to form amyloid plaques, are formed from the amyloid precursor protein (APP) after cleavage by BACE. Initial cleavage of APP by α-secretase rather than BACE prevents the eventual generation of amyloid-β, which would otherwise form the plaque which causes impaired brain function and the symptoms of Alzheimer's. Currently, research focused on finding inhibitors of the BACE to stop the formation of amyloid plaque is ongoing. The physiological purpose of BACE is unclear, but will be a focus of Alzheimer's research in the future.

To conclude, in order to overcome many of the challenges that medicine faces in the future, medical research will be reliant on the work of medicinal chemists to develop drugs that will help to treat the problems caused by diseases that current therapies fail to address. These solutions, just like the pioneering work on drug discovery in the past, will be governed by the fundamental principles of chemistry. Foundational to organic chemistry, the construct of all living organisms is based on the element carbon. Atoms of this element are bonded together by sharing pairs of electrons to build the scaffold onto which other elements essential to organic chemistry, most notably hydrogen, oxygen and nitrogen, can bond to create the macromolecules, such as proteins, nucleic acids, carbohydrates and lipids, which comprise the materials for building the cell. Biochemists study cells and the molecules of which they are composed to understand how cellular processes operate and relate to the function of the body, including the occurrence of disease. By understanding the molecular basis of disease, chemists are able to intelligently design drugs that will influence the biology of the cell in a way that alleviates the symptoms of the disease. The product of a century of scientific ingenuity, medicines are available today that would have seemed miraculous one hundred years ago. It is worth acknowledging, though, that the challenges presented to us by diseases in the future will require advancements in the field of medicinal chemistry that will be perceived as equally extraordinary.

Index

Note: *Italicized* page numbers refer to figures